WOMEN

WOMEN
Alcohol and Other Drugs

Edited by
Ruth C. Engs, R.N., Ed.D.

Alcohol Drug Problems Association
444 N. Capital Street, N.W.
Washington, D.C.

Resource *Publications*
An imprint of *Wipf and Stock Publishers*
199 West 8th Avenue • Eugene OR 97401

Resource Publications
A division of Wipf and Stock Publishers
199 W 8th Ave, Suite 3
Eugene, OR 97401

Women: Alcohol and Other Drugs
By Engs, Ruth C.
Copyright©1990 by Engs, Ruth C.
ISBN 13: 978-1-59752-887-0
ISBN: 1-59752-887-0
Publication date 8/22/2006
Previously published by Kendall/Hunt Publishing Company, 1990

CONTENTS

Foreword, ix
Preface, xi
Principle Article Authors, xiii

PART A: INTRODUCTION AND OVERVIEW   1

Chapter 1
   Understanding the Issues: An Overview *by Lois R. Chatham,* 3

PART B: PSYCHOSOCIAL ASPECTS   15

Chapter 2
   Alcohol Abuse and Alcoholism: Extent of the Problem *by Sharon C. Wilsnack,* 17

Chapter 3
   Etiology of Alcohol and Other Drug Problems: Nature vs Nurture *by Karol L. Kumpfer, Ann Holman Prazza and Henry O. Whiteside,* 31

PART C: PHYSIOLOGICAL ASPECTS   41

Chapter 4
   Alcohol and Hormones: Reproductive and Postmenopausal Years *by Judith S. Gavaler,* 43

Chapter 5
   Physiological Effects of Cocaine, Heroin and Methadone *by Janet L. Mitchell and Gina Brown,* 53

Chapter 6
   Alcohol, Pregnancy, and Fetal Development *by Lyn Weiner and Barbara A. Morse,* 61

PART D: PREVENTION AND INTERVENTION 69

Chapter 7
Prevention Issues in Developing Programs *by Darlind J. Davis,* **71**

Chapter 8
Employee Assistance Programs *by Mary Ellen Lukina-Wiersma,* **79**

Chapter 9
Adolescent Women *by Gail Gleason Milgram,* **85**

Chapter 10
College Women *by Ruth C. Engs,* **93**

PART E: TREATMENT 101

Chapter 11
Issues in Alcoholism Treatment *by Jacqueline Wallen,* **103**

Chapter 12
Opiates *by Marsha Rosenbaum and Sheigla Murphy,* **111**

Chapter 13
Cocaine *by Barbara Lynn Eisenstadt,* **119**

PART F: SPECIAL ISSUES 125

Chapter 14
Black Women *by Maxine Womble,* **127**

Chapter 15
Mexican-American Women *by Juana Mora,* **137**

Chapter 16
Native American Women *by Candace M. Fleming and Spero M. Manson,* **143**

Chapter 17
  Lesbian Women *by Dana Finnegan and Emily McNally,* **149**

Chapter 18
  Co-Dependency and Dysfunctional Family Systems
  by Sharon Wegscheider-Cruse, **157**

Chapter 19
  Domestic Violence *by N. Ann Lowrance,* **165**

FOREWORD

This publication of Women: Alcohol and Other Drugs is another contribution by the Alcohol and Drug Problems Association (ADPA) to the understanding of and in response to gender differences in addictions. Over a decade ago the membership of the ADPA formed a Women's Commission. This leadership group within the Association worked diligently to infuse the ADPA educational events with presentations which examined the issues of gender in the treatment of addiction. The Women's Commission also addressed the issues of professionals working in programs which were oriented toward the treatment of males.

Since 1987, ADPA and its Women's Commission has presented a national conference on gender issues annually. These conferences demonstrated that a substantial body of knowledge had developed during the last decade. The Association on behalf of the Women's Commission, asked the J. M. Foundation to support the development and editing of a publication which would present the knowledge and skills which have emerged about gender issues during the last decade. The J. M. Foundation presented ADPA with a grant award which permitted us to proceed.

All the contributors to this effort volunteered their talent and knowledge. Dr. Ruth Engs, the editor, was the person who put the project together. The authors gave freely of their time and talent. This has been a project marked by commitment and cooperation. Within this framework, chapters were produced on time, the editing was completed, and publication accomplished. My congratulations and thanks to all who participated.

A note of thanks to our publisher. It was ADPA's intention from the beginning of the project to offer the book to the field at a modest cost. It was the technical knowledge and willingness of the Kendall/Hunt Publishing Company to work within the time and fiscal constraints of ADPA which ultimately accomplished the production of this volume in August of 1989.

The ADPA Women's Commission and ADPA will continue to address the gender issues of addiction. We are pleased to present this book as a tool for understanding and a guide for more effective treatment of women by the programs and professionals of our field. With the sustained efforts of the ADPA Women's Commission and its commitment to the annual Women's Issues Conference, the Association intends to continue its contribution to the understanding of and an effective response to the gender differences of addicted persons.

> Karst J. Besteman
> Executive Director,
> Alcohol and Drug Problems Assocation
> August 1989

PREFACE

This monograph is a sample of some thoughts and trends concerning issues involving women and alcohol and other drugs during the year of ADPA's 40th anniversary. It is not meant to be inclusive of all aspects of the problem, but rather a look at current trends, thoughts and philosophies in the field. Comments and opinions are the authors' and do not necessarily reflect the opinions and philosophies of ADPA, the publisher, or the editor.

Authors were chosen to represent divergent geographic areas in the United States, various professional backgrounds and places of employment, diverse philosophies and opinions, and special or topical issues to get the broadest sampling of what is current and emerging in the field. Some authors are well known in their area of expertise and others are just beginning to emerge as leaders.

Over the past 40 years women, as professionals, have entered the addiction field and the authors of this monograph reflect this current leadership. All aspects of the field including research, intervention, treatment, prevention and education at the community, academic, government, public and private sectors, and at all levels of involvement, are represented in this publication. Forty years ago a monograph concerning women's alcohol and other drug issues, in which all the principle authors were women, would not have been possible. Also, at this point of time, women were not seriously considered, by many, to be at risk for alcohol or other drug abuse problems or to be acutely affected by alcohol and drug abusing family members. This monograph reflects the emergence of women over the past 40 years, not only as professionals in the field, but, also, as a population which has finally been recognized as having problems related to alcohol and other drugs.

This monograph has information useful to professionals who are involved with prevention, education, intervention, treatment, public policy and management. It has material which will serve as a basis or stimulus for further research and scientific investigations in the field. Furthermore, it has information which lay persons, or anyone interested in women's issues will find provoking.

The editor hopes the reader will find this monograph an inspiration for prevention, intervention and treatment of alcohol and other drug abuse problems, and related issues among women.

Appreciation is expressed to Lois R. Chatham for excellent editorial comments and suggestions which greatly enhanced the scope and quality of this publication.

Ruth C. Engs, Editor
April 1989

PRINCIPLE ARTICLE AUTHORS

*Lois Chatham
Former Director
Division of Basic Research, NIAAA
1612 Wiltshire Court
Roanoke, TX 76262

*Darlind Davis
Deputy Director
Division of Prevention and
 Implementation, OSAP
5600 Fishers Lane
Rockville, MD 20857

*Barbara Eisenstadt
Director, Substance Abuse
 and Cocaine Programs
Mediplex, Conifer Park
Scotia, NY 12302

*Ruth Engs
Associate Professor
Department Applied Science, HPER
Indiana University
Bloomington, IN 47405

*Dana Finnegan and Emily
 McNally
Co-Directors, Discovery
 Counseling Center
708 Greenwich St. 6D
New York, NY 10014

*Candace M. Fleming and
 Spero Manson
National Center for American
 Indian and Alaska Native
 Mental Health Research
Dept. of Psychiatry,
 University of Colorado
 Health Science Center,
Denver, CO

*Judith Gavaler
Associate Professor
School of Medicine
1000 J. Scaife Hall
University of Pittsburg
Pittsburg, PA 15261

*Karol Kumpfer, Ann Holman
 Prazza and Henry
 Whiteside
Director ALTA Institute
Social Research Institute
University of Utah
Salt Lake City, UT 84112

*N. Ann Lowrance
Director, Domestic Violence
 Service
Oklahoma Department of
 Mental Health
P.O. Box 53277 Capital
 Station
Oklahoma City, OK 73152

*Principle author and her address

*Mary Ellen Lukina-Wiersma
Regional Supervisor
Hazeden Services
26911 N.W. Highway,
  Suite 300
Southfield, MI 48034

*Gail Milgram
Director, Education and
  Training
Center for Alcohol Studies
Rutgers University
Piscataway, NJ 08854

*Janet Mitchell and
  Gina Brown
Obstetrics/Gynecology
Harlem Hospital
506 Lenox Avenue
New York, NY 10037

*Juana Mora
Planning Division
Office of Alcohol Programs
714 West Olympic, 10th Floor
Los Angeles, CA 90042

*Marsha Rosenbaum and
  Sheigla Murphy
Director, Institute for
  Scientific Analysis
2235 Lombard St.
San Francisco, CA 94123

*Jacqueline Wallen
Staff, Treatment Research
  Branch
NIAAA, Room 16CO3
5600 Fishers Lane
Rockville, MD 20857

*Lyn Weiner and
  Barbara Morse
Director, Fetal Alcohol
  Project
7 Kent St.
Brookline, MA 02146

*Sharon Wilsnack
Professor, Psychiatry
University of North Dakota
Grand Forks, ND 58201

*Maxine Womble
Chairperson, Board of
  Directors
National Black Alcoholism
  Council
417 S. Dearborn St.
Chicago, IL 60695

*Sharon Weigscheider-Cruse
Director, On Site
2820 West Main St.
Rapid City, SD 57702

*Principle author and her address

Part A

INTRODUCTION AND OVERVIEW

CHAPTER 1

Understanding the Issues
An Overview
Lois R. Chatham, Ph.D.

PURPOSE AND PROCESS

The purpose of this monograph is to record in a single document a variety of issues currently being addressed by those active in the field of women, alcohol and drug abuse. The authors include clinicians, scientists, university professors, government employees, and representatives of the private sector. A number of the authors identified themselves as members of a minority constituency.

Because the purpose of this monograph is to capture a broad sample of viewpoints and research findings from those currently active in the field of women, alcohol, and drug abuse, the findings and opinions presented by the authors have not been altered. As a result, readers may find one author's ideas inconsistent with another's. Moreover, some of the ideas expressed are controversial. Although the ideas expressed in this monograph do not necessarily represent the position of the Alcohol and Drug Problems Association (ADPA) on a given issue, the existence of the monograph does reflect ADPA's commitment to disseminate current topical information and to encourage dialogue about important issues within the field of alcohol and drug abuse.

HISTORICAL PERSPECTIVE

The findings and views contained in this 1989 monograph should be viewed within the framework of history. To do this, a few historical facts will be recalled for the reader. Although accounts of heavy alcohol consumption by Colonial Americans is fairly well documented, and evidence exists that some women as well as men engaged in heavy drinking, the issue of alcohol, as a social problem, did not gain historical importance until the Temperance Movement of the 19th and 20th centuries. During that period, men were the targets of the Movement because of their perceived excessive use of alcohol and their frequenting of saloons. During this period, bar room songs and artists' sketches depicted women and children as the victims, especially economically, of the alcohol excesses of male members of the family.

Women were important actors in the Prohibition Period which followed the Temperance Movement where history records that a sizeable number of them become Suffragettes in order to gain the right to vote, for the primary purpose of prohibiting the sale of alcohol which they believed was ruining many homes.

The passage of the Harrison Act in 1914 set the stage for viewing women who used opiates as criminals. The fact that many women addicts resorted to prostitution and other crimes to support themselves and their drug habits served to reinforce the view that all women with drug abuse problems were undesirable persons. Indeed it was not until the Nixon Administration, in an attempt to control crime, pushed Congress to increase direct federal support of an expanded heroin treatment system, that increased treatment for drug abusing persons, including women, was made available. And while the majority of the emphasis of the Nixon Administration was directed toward the treatment of persons using illegal drugs, funds for the treatment of alcohol abuse also increased modestly. With this increase in federal support, initially women were provided care as part of mixed gender programming; however, with more experience and more funding, many programs gradually expanded to develop specialized treatment programs for women.

In the late 1970's and early 1980's, as the end of the era of direct federal support of treatment programs neared, a group of lobbyists emerged who attempted to persuade Congress to increase grant funds for the support of specialized women's treatment programs. As a result, when the responsibility for treatment programs shifted from the federal government to the States, through the block grant legislation, these lobbyists continued to advocate for specialized treatment programs for women. The success of these efforts is clearly demonstrated by the continuation of block grant legislation which ear marks funds for the treatment of women.

During the 1980's another lobby group, Mothers Against Drunk Driving (MADD), became a powerful force for legislation and judicial reform aimed at protecting innocent drivers from those who drive while intoxicated. This advocacy group tended to concentrate its efforts at the state and local level. During this same period an unexpected objection was made by feminist groups to the posting of signs in New York City bars advising women that alcohol consumption could result in the birth of a child with Fetal Alcohol Syndrome (FAS). Although the group's objection to holding women solely responsible for FAS was scientifically premature, recent research suggests that the heavy drinking father may, indeed, be a contributing factor to the birth of a child with FAS.

Throughout history, women have taken an active, often aggressive, role in trying to reduce the damage done by the consumption of drugs and alcohol to members of their families and to others. And although in 1943 a woman recovering alcoholic, Marty Mann, gained public visibility by founding the National Council Alcoholism, only recently has alcohol and drug abuse as a woman's health issue undoubtedly has been accelerated by the open dialogue established with the news media and the public by the wives of two nationally prominent politicians.

OVERVIEW[1]

Size of the Problem

Estimating the size of the problem of alcohol and other drug abuse among females in the U.S. population is admittedly imprecise. However, recent national and special surveys permit reasonably accurate estimates of the current prevalence of alcohol abuse and alcoholism among adult women. In addition, surveys of the use of alcohol and other drugs by high school and college age students permit the estimation of the problem among in-school samples of young women. Within these constraints, best estimates of the present size of the problem are described below:

Adult Women (Age 18 and Older)

Indicators consistently point to the fact that alcohol and drug abuse is a serious problem among American women today. Of the 17.6 million Americans experiencing significant consequences associated with drinking, approximately one third (5.7 million) are women. Of these, 3.3 million are *alcoholics* and 2.4 million are *alcohol abusers*.[2] Relative to the total population, this suggests that one woman in sixteen is an alcohol abuser or alcoholic and that one of every ten women who drinks (40% of women are abstainers) have alcohol related problems.

Drinking patterns and drinking problems in women show rather consistent relationships to a variety of socio-demographic characteristics. Younger women (age 18–34) show higher rates of drinking-related problems than do women in older age groups, while middle aged women (35–49) show higher rates of chronic alcohol dependence or alcoholism. In terms of socio-economic status, rates of drinking (nonabstention) tend to be higher among women with higher levels of educa-

---

1. This overview is based primarily on the material presented by the authors of this monograph. Statements are sometimes lifted verbatim and sometimes paraphrased. Readers should go to the original chapters for reference citations.

2. See Chapter 2 for definitions of these two terms.

tion and income, but rates of drinking problems and alcohol dependence symptoms are not consistently related to socio-economic level. However, ethnic based differences in drinking patterns and health consequences have been found.

Higher rates of heavy drinking and drinking problems are found among divorced, separated or never married women than among those who are married or widowed; and cohabiting women are more likely than other women to drink heavily and experience drinking problems.

Based on the best available data it is projected that the number of U.S. adult women alcohol abusers and alcoholics will increase by nearly 200,000 women between 1985 and 1995 to an estimated total of nearly 6 million women in 1995. In absolute numbers, the largest age group of alcohol abusers and alcoholics will continue to be the 21–34 age group. In rate of change, the projected increase will be greatest among women aged 35–49, from 1.5 million in 1985 to more than 2 million in 1995, a 36% increase. This marked increase reflects the aging of the baby boom generation into the 35–49 age range where rates of chronic alcohol problems among women are highest. In addition to alcohol use, data from a major 1985 national household survey indicates that 1/2 of 1% of women reported heroine use in their lifetime; 27% of women reported a lifetime use of marijuana; and 8% of women reported lifetime use of cocaine, with much greater percentages of younger, than older, women reporting such use.

Females in College

Studies consistently show that female college students are less likely to drink and to get into problems with their drinking than are male students. However, college women have been drinking more frequently and in greater quantities since the late 1970's compared to the 1950's and 1960s. Findings from the 1988 survey of *Illicit Drug Use, Smoking, and Drinking by America's High School Students, College Students, and Young Adults* found that 90% of women in college identified themselves as having consumed alcohol at least once during the past year, 37% reported having used marijuana, 15% reported having used cocaine and 11% reported having used stimulants other than cocaine. Data from a more select sample of college students enrolled in 56 colleges throughout the U.S. shows a decrease in the percent of females students drinking once a year or more since 1985 (80% in 1985 vs 73% in 1988). This downward trend, found also in college males appears to be the result of raising the drinking purchase laws. This college survey also found that this decrease in drinkers has not resulted in a change in the percentage of *heavy* drinking among college women over the decade nor in the number of alcohol abuse behaviors[3]—other than drinking and driving problems which have continued to decrease since 1982. The discrepancy between these two surveys probably reflects the difference in statistical procedures and definitions of drinking. Both surveys clearly indicate there is extensive use of alcohol and others drugs by college women.

Females in High School

Surveys of adolescents consistently find that the majority of high school students drink, and that boys drink more frequently and drink larger amounts than do girls. A 1987, national survey found that 91% of the female high school students described themselves as drinkers. Of greater concern than the percentage who report drinking is the fact that of the females who reported drinking, 3% drink daily (and therefore may be dependent on alcohol), 34% on occasion drink to intoxication (thus placing themselves in high risk situations) and 9% indicate they drive after drinking. In addition, 54% of the 16–18 year old females indicate they have been a passenger in a car with an intoxicated driver. And although adolescents say they drink for the same reasons adults do, (to relax, be social, or have a good time), drinking to intoxication is a more accepted outcome of a drinking experience for adolescents than for adults. In addition to drinking alcohol, the same survey revealed that 48% of the high school senior girls reported some lifetime use of marijuana, 23% the use of stimulants, 14% the use of cocaine, and 14% the use of inhalants.

Nature of the Problem

The nature of the problem of alcohol and drug abuse among women is complex. Recent research findings indicate that both environmental and genetic factors play important roles in the emergence of these problems. Interest in the *environmental* causes and correlates of alcohol and drug abuse in women has grown since 1970; however more theories than facts abound. Some feel that increased emphasis on sexual equality and social acceptance of drinking and smoking in women may at least in part account for the increase in such behaviors. Whatever the causes, international studies report increasing alcohol, tobacco and drug use in women throughout the world.

In an attempt to account for perceived increases in drug use, some have observed that tobacco and alcohol advertisers have been quite successful in associating the use of alcohol and tobacco with beauty, wealth, social popularity and financial success; others speculate that the

---

3. See Chapter 10 for description of alcohol abuse behaviors.

value American society places on thinness, high energy and sexuality may be factors in inducing women to use both tobacco and cocaine.

In an attempt to discourage IV drug use and the spread of AIDS, considerable efforts are being exerted to persuade women that they play an important role in controlling the spread of this disease through their sex and drug use practices. Publicity about the role alcohol and other drugs have on a woman's health and that of her fetus, also are part of the environment in which drug taking by women now takes place.

Unfortunately, concern about health and wellness is at this time largely a middle and upper class phenomenon.

While the etiology of alcohol and other drug problems in women is unclear some research findings seem relevant. Women, more than men, attribute their alcohol and drug taking behavior to critical life events such as divorce or separation, death of a loved one, a dysfunctional childhood family unit, alcoholic or sexually abusive male family members and mental distress such as depression and anxiety. Women's initiation to drug use is usually by a male companion and more women than men alcoholics have alcoholic spouses. In light of these findings, it is not surprising that women occupy leadership positions in activities focusing on issues of codependency and in the Adult Children of Alcoholics movement.

On the *biologic* side, research is not definitive regarding the role that genetics plays in the development of alcohol and other drug abuse problems but it does give credence to the adage "like mother like daughter" in describing the genetic association which exists between alcoholic mothers and daughters. Strong associations also have been found between being the daughter of an alcoholic father and being diagnosed as suffering from depression or a somatoform illness in adulthood.

Although it is generally accepted that drugs have a profound effect on women's health, studies of drug effects on human female subjects have been limited to 1) difficulties presented by continual changes in hormonal status caused by the menstrual cycle and 2) concern over the ethics of administering drugs of any kind to women of child bearing age.

In spite of these constraints, research on animals and humans indicates that chronic exposure to high doses of ethanol or alcoholic beverages, as well as to cocaine and opiates clearly disrupts the menstrual cycle, inhibits ovulation (and thereby may adversely affect fertility) and may lead to an early menopause.

That alcohol has the potential for disrupting fetal development is well established. In spite of this knowledge, in the U.S. the incidence of Fetal Alcohol Syndrome (FAS), is 1.9 per thousand live births and it is likely that Fetal Alcohol Effects (FAE) occurs three times more frequently than FAS.[4] Children with FAS have been born only to women who drink abusively and a clear dose response effect on birth weight, mortality, and morphologic anomalies seem to exist. Research also indicates that the fetus has considerable capacity for compensatory growth and repair when the mother stops drinking. Recent animal and human studies suggest that the father's drinking behavior may be a contributing factor in FAS.

Similarly, animal as well as clinical studies on marijuana, cocaine, heroin and other illicitly used drugs indicate that material consumption of these substances during pregnancy is potentially damaging to the development of the fetus and may complicate the mother's health during pregnancy. In addition, the consumption of some classes of drugs results in the delivery of an addicted fetus. And although the administration of methadone during pregnancy has been studied extensively, the question of whether women can be safely detoxified from methadone during pregnancy remains controversial.

Historically, research in the area of alcohol consumption by postmenopausal women, where changing hormonal status would not be a compounding variable, concentrated on studying the effect of chronic, heavy alcohol consumption. As a result, research results frequently were difficult to interpret because the subjects manifest other health problems. Preliminary findings from one study looking at the effect of moderate alcohol use in healthy postmenopausal women suggest that alcohol consumption may result in changes in hormonal functioning known to mediate the risk of osteoporosis, cardiovascular disease and cancers of the breast and uterus.

The recent upsurge in the use of cocaine has resulted in the establishment of a relationship between the use of drug and cardiac arrhythmia, myocardial infarction and cerebrovascular accidents. And women who use street drugs run risks due to the presence of noxious diluents. Drug users who engage in intravenous and subcutaneous injections also suffer complications such as skin abscesses, cellulitis, and frequently fatal bacterial endocarditis and septic thrombophletibis. The most recent health problem to emerge for women who inject drugs is the risk of acquiring AIDS.

Prevention and Intervention in the Problem

Preventing women from using illegal drugs and inappropriate using alcohol is an often articulated public health goal. Ironically, for the past decade, in spite of legislation aimed at developing programs for women with drug abuse problems, with the exception of efforts targeted at

---

4. Chapter 6 for definitions of FAS and FAE.

women for the purpose of reducing the incidence of FAS, national prevention efforts have been overwhelmingly aimed at youth.

Conventional wisdom, reinforced by some documented success, says prevention strategies are most successful when the behavior which is to be prevented is easily identifiable, is perceived by the object of the prevention message as possible to do, and the benefit of the actions are seen as outweighing the costs. Using this model, prevention programs have been developed for the purpose of preventing alcohol related birth defects in populations of heavy drinking pregnant women. Research findings from several, independent investigators indicate that the majority of women, when advised to stop drinking during pregnancy do so, and that those who do not abstain, nevertheless, substantially reduce their alcohol intake. Examination of the incidence of FAS and FAE in the offspring of these subjects versus populations where such programs were not used, verifies the validity of this approach for both reducing fetal damage and for improving the nurturing ability of the mother for the newborn and for other children in her care.

Logic also suggests that changes in legislation, in advertising, in taxation and in the provision of educational material will reduce drug related problems. The hypothesis that such approaches would be useful was reinforced earlier this year when a nation-wide survey of a sample of college students revealed a significant decrease in the percentage of female drinkers which temporally coincided with changes in the legal age of purchase.

Consensus seems to exist that the goals of educational and intervention strategies targeted at females, whether introduced at home, in the school, in the physician's office or in the work setting, are to promote healthy behaviors based upon a full understanding of the effects and risks of drug use. There also is general agreement that to optimize effectiveness and to save costs prevention strategies should be targeted at high risk groups. And, although research now can identify which groups of females are at highest risk for developing alcohol and other drug related problems, knowledge about the effective methods of intervention is lacking. Therefore, at the present time, programs for reaching these target groups must, of necessity, be developed on the basis of professional judgment and conventional wisdom.

Treatment of the Problem

In spite of the large number of women experiencing alcohol and drug abuse problems, relatively few of them are in treatment. To be more precise, best estimates based on NDATUS and national prevalence data indicate that 1 out of 3 alcoholic or alcohol abusing persons in the U.S. is a women, and 1 of 20 of these women is in treatment in a given year. Of the other drug users, 2 of 5 is a woman and 1 of 50 of these is in treatment in a given year. National surveys of patients in treatment indicates that 24% of those in treatment for alcoholism are women and 33% of those in treatment for other drug abuse are women. In spite of the consistent call for specialized treatment programs for women, only 27% of treatment units offer such specialized programs and the proportion of treatment units with specialized women's programs is not increasing.

If so many women are suffering from alcohol or other drug abuse problems, why are they not in treatment? The most consistently mentioned barriers to coming in to treatment, given by women already in treatment, are the following. The woman, not unexpectedly, denies she has a problem. Less anticipated is the barrier caused by her family's denial that she has a problem. Sometimes this denial takes the form of the family's open opposition to a woman seeking help, simply because she is felt to be needed at home. The lack of family support for seeking treatment sometimes is complicated because other family members also have drug abuse problems. A consistent reason given by women for not seeking treatment is the lack of available child care resources. This can be an almost insurmountable barrier for many women whose absence from the home during treatment can jeopardize their already strained relationship with their families. Closely tied to this is the ever present threat of the loss of child custody. For some women the lack of insurance coverage is also a barrier, as is the failure of community gate keepers (physicians, social agencies and the clergy) to recognize alcohol and other drug problems. Important as identification of these barriers is, definitive knowledge about why women do not enter treatment will not be understood unless studies are conducted on those who have never reached treatment.

Little definitive information is available about how successful the treatment experience is for those women who enter treatment. The limited outcome data that is available suggests that women "do as well as men." However, it is not known what treatment regimes work best for women or whether specialized treatment programs are more effective than traditional programming. Consensus of those working in the field is that to be clinically relevant programming must acknowledge that: drug use patterns change with age; sensitivity to differences in ethnicity and sexual orientation is imperative if treatment is to be relevant; many women in treatment will have been victims of physical and sexual abuse at some time in their life; many of them will have eating disorders; a significant number of them will be diagnosed as suffering from anxiety and depression; and many of them will be dually addicted.

The treatment for addiction to heroin and cocaine warrant special attention because of their potential role in the spread of AIDS. For the female heroin addict, treatment is typically only a small component of a life long criminal career. When the female heroin addict enters treatment, it will most likely be in a methadone maintenance (MM) program in which methadone will be medically administered as a substitute for self administered heroin. This treatment approach is in sharp contrast to the treatment of other addictions in which total abstinence is prescribed. When successful, this treatment does away with the need to use drugs intravenously and thus reduces the risk for AIDS. At the present time MM is the largest single treatment modality for the female heroin addict. A legal substitute for cocaine is not available and those who treat women cocaine addicts find the addiction extremely frustrating. Current clinical practice consists of the use of therapies traditionally used with patients addicted to other non-opiate drugs. At present the female addicted to cocaine is a primary challenge to the field of public health because of the risk of contracting and spreading AIDS and the refractory nature of the illness.

Special Issues

With the recognition that alcohol and other drug abuse are legitimate women's concerns has come an awareness that before all women can benefit from what is known about the etiology, prevention and treatment of substance abuse problems some specific issues need to be addressed. Among the most talked about issues in the field today are those related to ethnicity, sexuality, victimization and co-dependency.

Alcohol and other drug abuse is felt by many to be one of the major health problems of the Black community. And although more *Black women* abstain than to Caucasian women, with increased socio-economic status and education Black women's drinking patterns tend to parallel those of the general female population. In addition, some health consequences of heavy drinking, such as cirrhosis and FAS, seem more severe in Black than Caucasian women. Similarly, poor, less well educated Black women are over represented in the population of women at risk for AIDS due to high risk drug and sex behaviors. Successful prevention and treatment programs for Black women must be color sensitive. To optimize success, such efforts should take into consideration the role that can be played by religion, professional and social organizations and the extended family.

Hispanic women report that they either abstain or drink very little. And although some researchers and health care professionals caution that cultural sanctions against female drinking may lead to under reporting, at the present time all indicators suggest that Hispanic women, as a group, do not suffer greatly from their own alcohol consumption or the use of other drugs. There is evidence, however, that they suffer disproportionately from the consequences of alcohol and drug abuse by male members of their families, including exposure to the risk of AIDS. Because the Hispanic population is the largest growing ethnic group in the U.S., even small changes in the drinking and drug taking patterns of Hispanic women could lead to a large number of affected persons. Programs for Hispanic women must be culturally relevant and viewed as safe, or they will not be used.

Although *American Indian* women are less likely to drink than the general U.S. population, a sample of drinkers in 20 American Indian communities found that 43.2% were women. Other studies further suggest that when controlling for the effects of social factors on drinking styles, the ratio of alcoholics to non-alcoholics was approximately the same for Indian men and Indian women. These findings confirm clinical impressions that in spite of the view that alcoholism is primarily a disease found in Indian males, it is disease that manifests itself in a significant number of Indian women as well. Congruent with this finding is the fact that alcoholism represents the fifth most frequent cause of death for American Indian women and contributes to four of the other ten leading causes of death. In addition to the harm alcohol consumption does to the woman, the prevalence of Fetal Alcohol Syndrome (F.A.S.) and Fetal Alcohol Effects (F.A.E.) are more common in many Indian communities than in the U.S. at large and reach unparalleled rates in some tribes of the southwestern Plains where FAS is the leading major birth defect. Although there is a fairly clear relationship between fewer years of formal education, lower occupational status and a higher preponderance of heavy drinking, little is known about the reasons why Indian women are at high risk for alcoholism and other alcohol related health consequences. Meanwhile, treatment and prevention programs which serve this population of women are attempting to make programs accessible, gender sensitive and culturally relevant.

Family violence and chemical dependency frequently co-exist. And although research has not established a causal relationship between the two events, those working in the field agree that when the conditions coexist, each behavior must be addressed to prevent injury—usually to a female. Studies show that 1/4 of the nation's homicides were committed by family members of the victim and that nearly half of these were between male and female partners. Of the partners killed, 2/3 were wives and 1/3 were husbands. The results of research conducted during the past ten years indicate that women who leave their batterers are at a 75% greater risk of being killed by the batterer than are those who stay. Since some studies show that 65–85% of women victims report alcohol was involved in the violent incident with a male spouse, (other drugs are reported in approximately 28% of the cases), and other studies suggest that the wives of batterers have about the same rates of alcohol and drug abuse problems as the general female population, the effects this fear of death or retribution has on a woman's willingness to seek treatment for alcohol and other drug abuse problems needs to be explored. In violent lesbian relationships, 35% of the batters had been under the influence of alcohol or drugs. Analyses of the dynamics of a variety of violent families reveals complex, co-dependent relationships.

Co-dependency, however, exists in many non-violent relationships. At present, it is a popular theory used to explain the personality dynamics of many Adult Children of Alcoholics. Since more men than women have problems with alcohol and drug abuse, it is logical that more women would become involved in such unhealthy relationships due to the drug dependency of a significant person or persons in their lives. The present ACOA focuses much of its efforts on helping women work through these unhealthy relationships. Since individuals involved in such family dynamics are at high risk for alcohol and drug abuse later in life, prevention efforts frequently are targeted to reach these individuals.

Issues of Sexuality are addressed in many contexts today. In the alcohol and drug abuse arena, sexuality, sexual identity, and sexual dysfunction are common topics of discussion. One of the most complex issues to address is that of the *lesbian* woman who has problems with alcohol and drugs. Although studies consistently show that the incidence and prevalence of alcohol and drug abuse problems is greater in the lesbian population than in the female population at large, no causal relationships have been established between sexual preference and problematic alcohol and drug use. Lesbians are not a special group, as are ethnic groups, but because of society's homophobic behavior, they are frequently treated as if they were and stereotypic behaviors are manifest in relating to them. One of the challenges facing the substance abuse field is finding ways to give visibility to this issue without reinforcing the view that lesbians are different from the general population. The resolution of this conflict can in part only be accomplished when society overcomes its timidity in addressing issues of sexual identity. When these issues can be handled openly, the stress that some lesbians ameliorate by the use of alcohol and other drugs may be eliminated. Meanwhile, prevention and treatment programs need to provide safe, non-homophobic environments in which lesbians, like all other women, can learn to live without drugs.

SUMMARY

In the past decade, the abuse of alcohol and other drugs by women have been recognized as important health issues. Studies consistently show the size of the problem to be significant although the magnitude and manifestation of the illnesses may vary relative to factors such as age, education, ethnicity, socio-economic status and community response. This monograph attempts to capture for the reader a cross section of the issues surrounding these illnesses in 1989.

Part B

PSYCHOSOCIAL ASPECTS

CHAPTER 2

Alcohol Abuse and Alcoholism Extent of the Problem[1]

Sharon C. Wilsnack, Ph.D.

Several recent national surveys permit estimates of the current prevalence of alcohol abuse and alcoholism among adult U.S. women. These include (1) a 1979 survey of 1,010 women and 762 men (Clark & Midanik, 1982); (2) a 1981 survey of 917 women and 396 men which included the largest national sample of heavier drinking women to date (R. Wilsnack et al., 1984); and (3) a 1984 survey of 3,128 women and 2,093 men which included larger samples of Black and Hispanic respondents than had earlier surveys (Hilton, 1987). National drinking surveys are typically designed to ask about drinking-related problems, not to make clinical diagnoses of alcoholism, and deciding what level of drinking problems indicates alcohol "abuse" is always somewhat arbitrary. (The Epidemiological Catchment Area Program of the National Institute of Mental Health (Regier et al., 1984) has recently adapted clinical diagnostic measures for use in general population surveys in selected sites, but these have not yet been used in representative national samples.) For these reasons, numbers of alcohol abusers and alcoholics derived from national drinking surveys should be regarded as the best available estimates of the extent of alcohol abuse and alcoholism in the general population, rather than as precise "head counts" of clinical "cases" of alcoholism.

NUMBERS OF ALCOHOL ABUSERS AND ALCOHOLICS

Using data on drinking problems from the 1979 national survey, the National Institute on Alcohol Abuse and Alcoholism (NIAAA) estimates that in 1985 there were approximately 3.3 million women alcoholics and an additional 2.4 million women alcohol abusers in the U.S. adult (over

---

[1]. Paper based upon testimony prepared for presentation at the Hearing on Alcoholism and Alcohol-Related Problems among Women, United States Senate Committee on Labor and Human Resources, Subcommittee on Children, Family, Drugs and Alcoholism, Washington, D.C., September 29, 1988.

18) population (Williams et al., 1987). *Alcoholics* are defined here as drinkers who experienced in the past year one or more symptoms of alcohol dependence (alcohol withdrawal symptoms or loss of control over drinking). *Alcohol abusers* are drinkers who during the previous year experienced at least one severe or moderately severe consequence of alcohol use, such as illness, job loss, or accidents, but who do not meet the criteria for alcohol dependence. This seems that a total of approximately 5.7 million American women are presently experiencing significant problem consequences associated with their drinking.

There are several ways of gaining perspective on these numerical estimates. Women make up approximately *one-third* of the *total number* of alcohol abusers and alcoholics (male and female) in the U.S., estimated at 17.6 million in 1985. Relative to the total population of *women*, the NIAAA estimates suggest that one woman in 16 is an alcohol abuser or alcoholic; relative to women *drinkers* (approximately 60% of U.S. women drink at least occasionally), *one woman drinker in 10* is an alcohol abuser or alcoholic. Similar rates were found in the 1981 national survey (R. Wilsnack et al., 1984), where one in nine women drinkers reported two or more drinking-related problems in the past year, and one in six women drinkers reported one or more symptoms of alcohol dependence.

The magnitude of alcohol abuse and alcoholism in women can be appreciated by viewing alcohol problems in relation to other major health problems of women. For example, alcohol abuse and alcoholism affect nearly two-thirds as many women as do chronic heart conditions (9.2 million, National Center for Health Statistics, 1988). More than 20 times as many women suffer from alcohol abuse and alcoholism as die annually from heart disease (250,000), from all forms of cancer (214,000), or from breast cancer (40,000) (NCHS, 1988). The economic and family roles of women, amplify the deleterious effects of women's alcohol problems far beyond the harm to women alcohol abusers themselves. In addition, the total domain of "women affected by alcohol abuse and alcoholism" also includes substantial numbers of women who experience negative consequences from *others'* drinking, for example, as spouses or partners of alcoholic men, victims of alcohol-involved traffic accidents, or victims of alcohol-related sexual or other abuse (Fillmore, 1985).

SPECIFIC ALCOHOL PROBLEMS

Women experience a wide range of specific alcohol-related problems that are not always reflected in overall estimates of alcohol abuse and alcoholism. While alcohol-related *traffic accidents and fatalities* are more common among men than among women, the number of female drivers in fatal traffic accidents increased 37% between 1977 and 1985; the number of male drivers in fatal accidents increased during the same period (Zobeck et al., 1987).

Although *liver cirrhosis* death rates have declined among both women and men in recent years, nearly 10,000 women in this country died of cirrhosis in 1985 (Grant et al., 1988); alcohol abuse is the leading cause of cirrhosis. Women appear to be more susceptible to cirrhosis than men at the same levels of alcohol consumption; alcoholic women develop cirrhosis at an earlier stage in their drinking careers and at lower levels of consumption than do alcoholic men (Hill, 1984). Cirrhosis deaths among nonwhite women continue to exceed those among white women, with American Indian women showing considerably higher rates than either black or white women (Bertolucci et al., 1985).

Alcohol abuse during pregnancy results in *fetal alcohol syndrome* (FAS) in an estimated 1 to 3 cases per 1,000 live births in the general population of women, and 23 to 29 cases per 1,000 live births among alcohol abusers (NIAAA, 1987). Applying an FAS rate of 1.9/1,000 live births (Abel & Sokol, 1987) to the 3.8 million live births in 1986 (NCHS, 1986) yields more than 7,000 FAS births per year. In addition, even larger numbers of infants are affected each year by *fetal alcohol effects* of varying degrees of severity.

Several recent studies have found associations between *obstetric and gynecologic problems* and heavy drinking in women (Russell & Coviello, 1988; S. Wilsnack et al., 1984). At present it is difficult to separate reproductive problems that *result from* heavy alcohol use from reproductive problems that *precede* and possibly contribute to women's heavy drinking. However, given that American women make an estimated 55 million office visits per year to obstetricians and gynecologists (Russell & Coviello, 1988), the numbers of women experiencing reproductive dysfunction caused or made worse by alcohol—and the associated costs of medical treatment—are likely to be substantial.

Elevated rates of *sexual dysfunction* have also been reported among heavy drinking and problem drinking women, although again it is difficult to distinguish sexual dysfunction *caused* by drinking from dysfunction *contributing to* drinking (Wilsnack, 1984). Approximately 60% of women drinkers report feeling less sexually inhibited after drinking (Klassen & Wilsnack, 1986), and heavier drinking in women is associated with nontraditional sexual behavior (such as premarital sexual relations) and with cohabiting. Research is needed to address possible implications of linkages between drinking and reduced sexual inhibition or restraint for the spread of the AIDS epidemic in the heterosexual population, in terms of relationships between drinking and safe-sex practices and the balance of female and male responsibilities for safe-sex behavior.

Demographic Variations

As in men, drinking patterns and drinking problems in women show rather consistent relationships to a variety of sociodemographic characteristics. *Younger* women (age 18–34) show higher rates of *drinking-related problems* than do women in older age groups (Hilton, 1987; R. Wilsnack et al., 1984), while middle-aged women (35–49) show higher rates of *chronic alcohol dependence or alcoholism* (Williams et al., 1987). In terms of *socioeconomic status,* rates of drinking (non-abstention) tend to be higher among women at higher levels of education and income, but rates of drinking *problems* and alcohol dependence symptoms are not consistently related to socioeconomic level.

As in earlier surveys, black women in the 1980s have higher rates of *abstention* than do white women. However, the tendency of black women *drinkers* to drink more heavily than white women drinkers, reported in earlier surveys, appears to have weakened or disappeared in the 1980s (Herd, 1988; R. Wilsnack et al., 1984). Data from a large subsample of Hispanic respondents in the 1984 national survey indicate that Hispanic women are considerably more likely to be abstainers and less likely to be heavy drinkers than are either non-Hispanic women or Hispanic men (Hilton, forthcoming). There are at present no data from representative national samples of American Indian women, but available data suggest that Indian women experience higher rates of fetal alcohol syndrome than white women, and higher rates of alcohol-related deaths, including death from liver cirrhosis, than either black or white women (Bertolucci et al., 1985; Leland, 1984).

Most surveys have found higher rates of heavy drinking and drinking problems among women who are *divorced or separated,* or who have *never married,* than among women who are married or widowed. In the 1981 survey *cohabiting* women were more likely than other women to drink heavily and experience drinking problems, and women who entered a cohabiting relationship between 1981 and 1986 showed increased rates of problem drinking in 1986 (Wilsnack et al, 1988).

There has been little recent evidence for the once-popular belief that women with multiple roles (e.g., married women with paid employment outside the home) are at especially high risk for alcohol abuse. In fact, women who are combining marital and employment roles may have *lower* rates of alcohol problems than women who *lack* these roles, i.e., who are unmarried, unemployed, or employed parttime (Wilsnack, et al., 1986). "Role deprivation"—a lack of meaningful social roles—

may increase women's risk for unrestrained or self-medicative use of alcohol and/or other drugs (Wilsnack & Cheloha, 1987).

Changes

Drinking among women in general increased markedly between World War II and the early 1970s and then appears to have remained relatively stable through the 1970s to the mid-1980s. However, changes within specific *subgroups* of women have been reported in several studies. Among young women aged 21–34, for example, comparison of 11 U.S. national surveys between 1964 and 1984 shows that heavy drinking (two or more drinks per day) increased from 4% in 1964 to 7% in 1984 (Hilton, 1988). Interestingly, heavy drinking did *not* increase among the youngest women, aged 18–20. Increased heavy drinking in women aged 21–34 is consistent with Fillmore's (1984) finding that heavy-frequent drinking was more common among women who in 1979 were 21–29 than among women who were surveyed at that age in 1964 and 1967. Rates of drinking (non-abstention) have increased among women aged 50–64 in the past 20 years, from 51% in 1964 to 62% in 1984, but without an increase in heavy drinking (Hilton, 1988). These variations in drinking trends for different age groups illustrate why it is important to continue to monitor trends in women's drinking and drinking problems as these are affected by age and by historical changes.

Projected Alcohol Abuse and Alcoholism, 1985–1995

NIAAA has recently projected that the number of U.S. women alcohol abusers and alcoholics will increase by nearly 200,000 women between 1985 and 1995, to an estimated total of nearly 6 million women in 1995 (Williams et al., 1987). As shown in Figure 2.1, in absolute *numbers,* the largest age group of alcohol abusers and alcoholics will continue to be women aged 21–34 (2.9 million in 1995). In *rate of change,* however, the projected increase will be greatest among women aged 35–49, from 1.5 million in 1985 to more than 2 million in 1995, a 36% increase. This marked increase reflects the aging of the baby-boom generation into the 35–49 age group where rates of chronic alcohol problems among women are highest.

ALCOHOLIC WOMEN IN TREATMENT

Despite the large numbers of American women who suffer from alcohol abuse and alcoholism, and the increasing scientific knowledge about and public awareness of women's alcohol problems in recent years, women are still greatly *under-represented in* and seriously *under-served* by the alcoholism treatment system in this country.

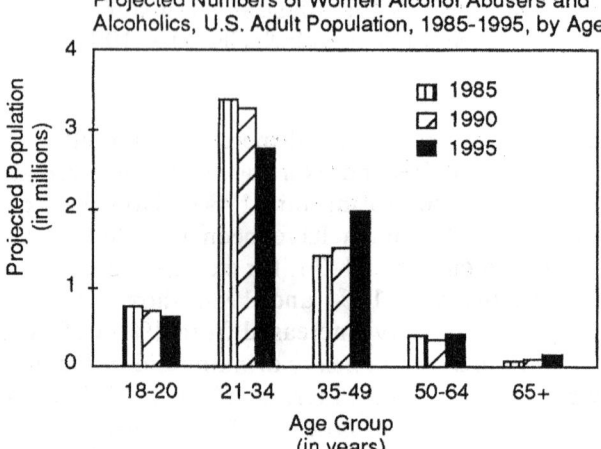

Figure 2.1 Projected Numbers of Women Alcohol Abusers and Alcoholics, U.S. Adult Population, 1985–1995, by Age.. (**Source:** Williams et al., 1987)

Women Clients in Alcohol Treatment

A 1987 government study collected information on treatment and client characteristics from 6,866 programs nationwide that provided alcoholism treatment, drug abuse treatment, or combined alcohol/drug treatment (NIDA/NIAAA, 1989). Of the 594,627 clients in alcohol or other drug abuse treatment at the time of the survey for whom data on client gender were available, only 164,712 (27.7%) were female. Women made up a higher proportion (32.8%) of the 257,750 clients with a primary diagnosis of drug abuse/dependence than of the 336,877 alcohol abuse/dependence clients (23.7% female). The proportion of alcohol clients who were women was well below the estimated proportion of alcohol abusers and alcoholics in the general population who are women—approximately one in three (33%). The findings from this national study are quite similar to those of a 1987 survey of all *state-supported* alcohol and drug treatment programs, which found that 19.8% of alcohol clients and 31.2% of drug clients were female (Butynski & Canova, 1988).

Specialized Services

A nationwide survey in 1984 of nearly 7,000 treatment units (Reed & Sanchez, 1986) collected information on services provided but not on characteristics of clients. Only a small minority of treatment units reported offering any specialized programs for women: 21.6% in 1984, down from 23.0% in 1982. The 1987 national treatment survey found that a slightly higher proportion (28%) of the 6,866 treatment units responding to a question about specialized programs reported providing at least some specialized services for women. A larger proportion of alcohol-only units (30%) than of drug abuse-only units (23%) or combined units (28%) provide specialized services for women.

Barriers to Treatment

A recent survey of more than 2,500 local alcohol authorities, alcohol treatment centers, and community gatekeepers (Association of Junior Leagues, 1988) asked respondents what they saw as the most serious barriers to women's seeking treatment for alcohol problems. All three respondent groups identified the same three barriers as most serious: (1) *personal denial* of alcohol problems by women themselves; (2) *responsibility for care of dependent children;* and (3) *family denial* of women's alcohol problems and/or family opposition to treatment. In the same survey, all three respondent groups identified treatment programs that provided *childcare* as the most needed resource for women with alcohol problems. These findings suggest that efforts to increase women's access to alcohol treatment need to take into account the family roles and responsibilities of many problem drinking women: decreasing family denial, enhancing family support and cooperation, and providing childcare in conjunction with alcohol treatment may be important strategies for improving women's chances of getting help for their alcohol problems.

The Association of Junior Leagues survey identified another major barrier to treatment: lack of awareness of women's alcohol problems on the part of key community *gatekeepers*—individuals and agencies serving as potential sources of referral to treatment. Several gatekeeper groups (including social services personnel, clergy, and physicians) indicated rather low levels of perceived alcohol problems in their women clients, and only a minority (less than 36%) of local alcohol authorities were making any effort to educate gatekeeper groups about women and alcohol. These findings indicate a need to target community gatekeepers, including law enforcement officials, social service personnel, clergy,and health care providers, with programs that will heighten their awareness of alcohol problems specific to women and their knowledge of available and needed programs for women with alcohol-related problems.

TARGET GROUPS FOR PREVENTION AND TREATMENT

Findings of recent epidemiological research on women's drinking suggest a number of subgroups of women that may be particularly important targets for alcohol abuse prevention and treatment efforts. This section discusses several such groups that may merit special attention based on their high-risk status and/or other special characteristics.

Young Women (Age 21–34)

As indicated earlier, young women report the highest rates of specific drinking-related problems, such as driving while intoxicated, conflicts with friends or family, or job problems. There is some evidence that young women's elevated rates of drinking problems are related to their greater tendency to engage in episodes or "bursts" of heavy drinking (e.g., on weekends or at parties), rather than spacing similar amounts of consumption over longer periods of time as may be more characteristic of older women (Bradstock et al., 1988; Wilsnack et al.,1986). Many young women will "mature out" of these early drinking problems without developing chronic patterns of alcohol abuse. Nonetheless, problem drinking behavior among young women is a serious concern, in terms of both its social costs and the threats it poses to the health and well-being of young women.

Additional reasons for special attention to young women drinkers include the evidence presented earlier that (1) heavy drinking increased in women aged 21–34 between 1964 and 1984 (Hilton, 1988); (2) heavy-frequent drinking may be increasing among young employed women (Fillmore, 1984); and (3) alcohol-involved traffic fatalities are increasing among women (Zobeck et al., 1987), with younger women the most likely to drive after drinking. Women in this age group are more likely than older women to exhibit several other behaviors associated with heavier alcohol use, including cohabiting, nontraditional sexual behavior, and nontraditional gender-role behavior (Wilsnack, in press). Other alcohol problems most prevalent among younger women include problems related to combined use of alcohol and illegal drugs, and risks of fetal damage from alcohol and other drug use in women of childbearing age.

Prevention and treatment programs for younger women need to take into account special characteristics and needs of women in this age group, e.g., their pattern of heavy episodic drinking in social settings, their greater likelihood of using both alcohol and illegal drugs, or the needs of young mothers who find it difficult to leave their children in order to seek treatment for alcohol-related problems. More research is needed on women in this age group to determine what factors predict whether a young woman will mature out of youthful problem drinking or will develop more chronic patterns of alcohol abuse and alcoholism.

Women Age 35–49

Women in this age group have the highest rates of chronic alcohol problems. They are more likely than younger women to experience symptoms of alcohol dependence, such as alcohol withdrawal symptoms or loss of control over drinking (Williams et al., 1987). Many women in this age group undergo losses of major social roles, such as divorce, separation, or children's departure, that may increase their risk of alcohol abuse (Wilsnack & Cheloha, 1987). As discussed earlier, the number of women alcohol abusers and alcoholics in this age group will increase by approximately 36% between 1985 and 1995, according to NIAAA projections. The impact of alcohol problems on women in this age group extends to their occasional or career roles (where they may be at the peak of their occupational productivity) as well as to their central roles within their families and other social groups.

Given the expected increase in 35–49-year-old women alcohol abusers and alcoholics during the coming decade, this is a critical group on which to focus prevention and treatment efforts. Prevention and early intervention efforts might include workplace-based programs for employed women, and programs to increase social support or provide training for alternative roles for women experiencing role losses or transitions. As with younger women, prevention and treatment strategies for 35–49-year-old women should pay particular attention to the role of family responsibilities and family support or denial as these influence women's development of and recovery from alcohol abuse and alcoholism.

Other Subgroups at High Risk for Alcohol Abuse

Table 2.1 summarizes a number of demographic and other factors that recent research suggests may increase women's risk of alcohol abuse and alcoholism. All the variables in Table 2.1 were associated with elevated rates of heavy drinking, drinking-related problems, and/or alcohol dependence symptoms in women in one or more cross-sectional studies, where drinking behavior and other variables were measured at the *same* point in time. In addition, variables marked by an (*) or a (+) were found to *predict* increases in women's drinking or drinking problems over a five-year period (Wilsnack et al., 1988). Longitudinal research of this sort can help to identify subgroups of women at high risk for subsequent alcohol problems, groups which in turn can be targeted by prevention and early intervention efforts. As examples, the finding in Table 2.1 that unemployed and parttime employed women have elevated rates of alcohol abuse may point to the importance of ensuring access to meaningful employment roles for all women who wish them. The finding that women experiencing depression, sexual dysfunc-

TABLE 2.1. Subgroups of Women at High Risk for Alcohol Abuse. (**Source:** Wilsnack et al., 1986, 1988.)

Demographic Subgroups at High Risk for Alcohol Abuse

Marital Status
　+ Never-Married Women
　　Divorced or Separated Women
　* Cohabiting Women
Employment
　+ Unemployed Women Seeking Work
　* + Parttime Employed Women
Parental Status
　"Empty Nest" Mothers

Other Groups at High Risk for Alcohol Abuse

Women Experiencing . . .
　*+ Depression
　+ Sexual Dysfunction
　*+ Reproductive Disorders
Women With . . .
　+ Nontraditional Gender-Role Orientations
　* Nontraditional Sexual Behavior
　+ Heavy-Drinking Significant Others
+ Young Women (Age 21-34)
Daughters of Alcoholics

*Predicts *onset* of drinking problems
+ Predicts *continuation* or *worsening* of drinking problems

tion, and reproductive disorders are at risk for later alcohol abuse suggests that prevention efforts might attempt to reach such women through mental health centers, sex therapy clinics, or obstetrics-gynecology clinics—*before* alcohol abuse becomes a second major problem. Similarly, women living with heavy-drinking husbands or partners might be reached through agencies that treat their spouses, and offered targeted alcohol education, social support, and skills training. Continued research on biological and psychosocial risk factors for alcohol abuse in women will permit future prevention and early intervention efforts to be targeted more precisely to the specific characteristics and needs of high-risk subgroups of women.

SUMMARY AND RECOMMENDATIONS

Findings presented in this paper suggest four major recommendations for needed action. First, it is important to continue to *monitor trends in women's drinking and drinking problems,* including trends within *age groups.* Although there do not appear to have been major drinking changes among women *in general* over the past decade, there is some evidence of change within specific age groups, for example, the possibility of increased heavy drinking among women aged 21–34. Close monitoring of these trends over time will allow us to determine whether most young women will "mature out" of early problem drinking or whether they will carry high rates of alcohol abuse with them into their middle years, and whether changes will occur in any other age groups that might have implications for prevention, treatment, or policy.

Second, attempts to reduce alcohol abuse in women must include attention to the wide range of specific *alcohol-related problems* women experience, including employment and family problems caused or made worse by drinking-related accidents and fatalities (on the highway and in the home), fetal alcohol syndrome and other fetal alcohol effects, alcohol-related sexual and reproductive dysfunction, other health problems, and possible effects of alcohol as a disinhibitor of unsafe sexual behavior. Not all problem drinking women experience all of these alcohol-related problems, and different prevention approaches are needed for different alcohol problems. For example, a program to prevent alcohol-related *loss of productivity* among employed women might involve workplace-based interventions, while a program to prevent *fetal alcohol damage* might attempt to reach women of childbearing age via public education, dissemination of information in obstetrics-gynecology clinics, and incorporation of FAS information in school health curricula. Efforts to reduce the psychosocial effects of maternal alcoholism on *children* might include family counseling, evaluation and treatment of children's psychological or behavioral problems, and support groups for both mothers and children. A comprehensive program to prevent alcohol abuse in women will require innovative approaches to reducing a variety of specific alcohol problems now experienced by women.

A third, related recommendation is for expanded treatment and prevention efforts that are sensitive to the special characteristics and needs of *subgroups* of women. There is no single, typical "alcoholic woman." A 75-year-old widowed woman abusing alcohol and prescription drugs has different treatment needs than does a 20-year-old unemployed alcohol abuser who also uses cocaine and marijuana, or a 35-year-old divorced woman working fulltime to support her young children. As discussed previously, a common need of women in many subgroups is for treatment programs that provide *childcare,* allowing women to seek alcohol treatment without fears about the welfare or threatened loss of their children. Training of *community gatekeepers* to heighten their awareness of women's alcohol problems would also benefit women in numerous subgroups.

Finally, implementing the preceding recommendations will require *increased support for research on women and alcohol.* Increased support would enable researchers studying a wide range of alcohol-related topics (e.g., genetic and family environment risk factors, alcohol-induced damage to such organ systems as the liver or to the central nervous system, or new approaches to prevention and treatment) to more frequently include both female *and* male subjects in their investigations and thus contribute to our understanding of both similarities and differences between women and men in the causes and consequences of their alcohol problems. Decreased support for both basic and clinical research on alcohol use and abuse is a critical component of a comprehensive effort to reduce the enormous personal and social costs of alcohol problems in women.

REFERENCES

Abel, E. L., and Sokol, R. J. (1987). Incidence of fetal alcohol syndrome and economic impact of FAS-related anomalies. *Drug and Alcohol Dependence, 19,* 51–70.

Association of Junior Leagues (1988). *Summary of findings: WOMAN-TO-WOMAN Community Services Survey.* New York: Association of Junior Leagues.

Bertolucci, D., Noble, J., and Dufour, M. (1985). Alcohol-associated premature mortality—United States, 1980. *Morbidity and Mortality Weekly Report, 34,* 493–494.

Bradstock, K., Forman, M. R., Binkin, N. J., et al. (1988). Alcohol use and health behavior lifestyles among U. S. women: The Behavioral Risk Factor Surveys. *Addictive Behaviors, 13* (1), 61–71.

Butynski, W., and Canova, D. (August, 1988). *State resources and services related to alcohol and drug abuse problems, Fiscal Year 1987: An analysis of state alcohol and drug abuse profile data.* Washington, D.C.: National Association of State Alcohol and Drug Abuse Directors, Inc.

Clark, W. B., and Midanik, L. (1982). Alcohol use and alcohol problems among U. S. adults: Results of the 1979 national survey. In *Alcohol consumption and related problems* (pp. 3–52). NIAAA Alcohol and Health Monograph No. 1. DHHS Pub. No.(ADM)82-1190. Washington, D.C.: Government Printing Office.

Fillmore, K. J. (1984). "When angels fall": Women's drinking as cultural preoccupation and as reality. In S. C. Wilsnack and L. J. Beckman (Eds.), *Alcohol problems in women: Antecedents, consequences, and intervention* (pp. 7–36). New York: Guilford Press.

Fillmore, K. M. (1985). The social victims of drinking. *British Journal of Addiction, 80,* 307–314.

Grant, B. F., Zobeck, T. S., and Ng, M. C. (1988). *Surveillance Report No. 8: Liver cirrhosis mortality in the United States, 1971–85.* Rockville, MD: National Institute on Alcohol Abuse and Alcoholism, Division of Biometry and Epidemiology, Alcohol Epidemiologic Data System.

Herd, D. (1988). *Gender role and socio-economic differences in drinking behavior in black and white women: Results from a national survey.* Working paper, Alcohol Research Group, Berkeley, CA.

Hill, S. Y. (1984). Vulnerability to the biomedical consequences of alcoholism and alcohol-related problems among women. In S. C. Wilsnack and L. J. Beckman (Eds.), *Alcohol problems in women: Antecedents, consequences, and intervention* (pp. 121–154). New York: Guilford Press.

Hilton, M. E. (1987). Drinking patterns and drinking problems in 1984: Results from a general population survey. *Alcoholism: Clinical and Experimental Research, 11,* 167–175.

Hilton, M. E. (1988). Trends in U. S. drinking patterns: Further evidence from the past 20 years. *British Journal of Addiction, 83,* 269–278.

Hilton, M. E. (forthcoming). The demographic distribution of drinking patterns in 1984. *Drug and Alcohol Dependence.*

Klassen, A. D., and Wilsnack, S. C. (1986). Sexual experience and drinking among women in a U. S. national survey. *Archives of Sexual Behavior, 15,* 363–392.

Leland, J. (1984). Alcohol use and abuse in ethnic minority women. In S. C. Wilsnack and L. J. Beckman (Eds.), *Alcohol problems in women: Antecedents, consequences, and intervention* (pp.66–96). New York: Guilford Press.

National Center for Health Statistics (1988). *Vital statistics of the United States, 1985. Vol. 2: Mortality, Part A.* DHHS Pub. No. (PHS)88-1101. Washington, D. C.: Government Printing Office.

National Institute on Alcohol Abuse and Alcoholism (1986). *Women and alcohol: Health-related issues.* Research Monograph No. 16. DHHS Pub. No. (ADM)86-1139. Washington, D. C.: Government Printing Office.

National Institute on Alcohol Abuse and Alcoholism (1987). *Sixth special report to the U. S. Congress on alcohol and health.* DHHS Pub. No. (ADM)87-1519. Washington, D. C.: Government Printing Office.

National Institute on Drug Abuse, Division of Epidemiology and Statistical Analysis; and National Institute on Alcohol Abuse and Alcoholism, Division of Biometry and Epidemiology (February 9, 1989). *Highlights from the 1987 National Drug and Alcoholism Treatment Unit Survey (NDATUS).* Rockville, MD: NIDA/NIAAA.

Reed, P. G. (Ed.) (1983). *Comprehensive Report: Data from the September 30, 1982 National Drug and Alcoholism Treatment Utilization Survey (NDATUS).* Rockville, MD: National Institute on Alcohol Abuse and Alcoholism.

Reed, P. G., and Sanchez, D. S. (1986). *Characteristics of alcoholism services in the United States—1984: Data from the September 1984 National Alcoholism and Drug Abuse Program Inventory.* Rockville, MD: National Institute on Alcohol Abuse and Alcoholism, Division of Biometry and Epidemiology.

Regier, D. A., Myers, J. K., Kramer, J., et al. (1984). The NIMH Epidemiologic Catchment Area Program: Historical context, major objectives, and study population characteristics. *Archives of General Psychiatry, 41,* 934–941.

Russell, M., and Coviello, D. (1988). Heavy drinking and regular psychoactive drug use among gynecological outpatients. *Alcoholism: Clinical and Experimental Research, 12,* 400–406.

U. S. Senate, Committee on Labor and Public Welfare, 94th Congress, Second Session (1976). *Hearings before the Subcommittee on Alcoholism on examination of the special problems and unmet needs of women who abuse alcohol.* Washington, D. C.: Government Printing Office.

Williams, G. D., Stinson, F. S., Parker, D. A., Harford, T. C., and Noble, J. (1987). Epidemiologic Bulletin No. 15: Demographic trends, alcohol abuse and alcoholism, 1985–1995. *Alcohol Health and Research World, 11* (3), 80–83.

Wilsnack, R. W., and Cheloha, R. (1987). Women's roles and problem drinking across the lifespan. *Social Problems, 34,* 231–248.

Wilsnack, R. W., Wilsnack, S. C., and Klassen, A. D. (1984). Women's drinking and drinking problems: Patterns from a 1981 national survey. *American Journal of Public Health,* 74, 1231–1238.
Wilsnack, S. C. (1984). Drinking, sexuality, and sexual dysfunction in women. In S. C. Wilsnack and L. J. Beckman (Eds.), *Alcohol problems in women: Antecedents, consequences, and intervention* (pp. 189–227). New York: Guilford Press.
Wilsnack, S. C. (in press). Drinking and drinking problems in women: A U. S. longitudinal survey and some implications for prevention. In T. Loberg, G. A. Marlatt, W. R. Miller, and P. E. Nathan (Eds.), *Addictive Behaviors: Prevention and early intervention.* Amsterdam: Swets & Zeitlinger.
Wilsnack, S. C., and Beckman, L. J. (Eds.) (1984). *Alcohol problems in women: Antecedents, consequences, and intervention.* New York: Guilford Press.
Wilsnack, S. C., Klassen, A. D., & Wilsnack, R. W. (1984). Drinking and reproductive dysfunction among women in a 1981 national survey. *Alcoholism: Clinical and Experimental Research,* 8, 451–458.
Wilsnack, S. C., Schur, B. E., Klassen, A. D., & Wilsnack, R. W. (June, 1988). *Predictors of change in women's drinking: A five-year longitudinal analysis.* Paper presented at the Annual Meeting, Research Society on Alcoholism, Charleston, SC.
Wilsnack, S. C., Wilsnack, R. W., & Klassen, A. D. (1986). Epidemiological research on women's drinking, 1978—1984. In *Women and alcohol: Health-related issues* (pp. 1–68). NIAAA Research Monograph No. 16. DHHS Pub. No. (ADM)86-1139. Washington, D.C.: Government Printing Office.
Zobeck, T., Grant, B. F., Williams, G. D., and Bertolucci, D. (1987). *Surveillance Report No. 6: Trends in alcohol-related fatal traffic accidents, United States: 1977–1985.* Rockville, MD: National Institute on Alcohol Abuse and Alcoholism, Division of Biometry and Epidemiology, Alcohol Epidemiologic Data System.

CHAPTER 3

Etiology of Alcohol and Other Drug Problems: Nature vs Nurture

*Karol L. Kumpfer, Ph.D.,
Ann Holman Prazza, M.S.W
and Henry O. Whiteside, Ph.D.*

Although interest in the causes and correlates of alcohol and drug abuse in women has grown since 1970, there is still little etiological research on this topic. Possibly because of increased sexual equality and social acceptance of drinking and smoking in women, the percentage of women drinking and smoking has increase dramatically in the last fifty years, particularly in young women (Gritz, 1986; Rachel et al., 1975; Wechsler & McFadden, 1976). Likewise, international studies report increasing alcohol, tobacco and drug use in women throughout the world (U.N. Secretariat, 1987).

Tobacco and alcohol advertisers have been quite successful in "liberating" young women to use by glamorizing use through associations with beauty, wealth, and social popularity. Increased endorsement that "smoking cigarettes keeps your weight down" is also associated with heavy smoking in women (Charleston, 1984). Whatever the reasons, more U. S. high school and college women than men smoke occasionally or heavily (a half pack per day) (Johnston, et al., 1986). A positive decreasing trend has recently been noted in these same young women for daily marijuana, alcohol and cigarette use since 1978 or 1979 and illegal drug use since 1981 (Johnston, et al., 1986).

Increased information on the potential damage to infants due to fetal alcohol or drug syndrome or effect has raised the general public concern about substance use and abuse in women. In addition, intravenous (IV) drug abusing women have the potential of developing AIDS and transmitting AIDS to their infants. Special legislation block grant set-asides have been developed for women's substance abuse treatment and the recent Omnibus Anti-drug Abuse Bill of 1988 authorizes special funding for detection and treatment of drug-abusing pregnant women. Nevertheless, little research is underway on the causes of chemical dependency in women. Without such information, treatment and prevention programs will often fail to address the most important causes of abuse in women.

Freud claimed that, even after considerable study, he did not understand what women want. Similarly after reviewing the scant research available, this author can not answer with assurance the question: "Why do women abuse alcohol and drugs?" Gomberg (1976), given the paucity of data, also found it difficult to answer the more refined question: "Why do some women manifest alcoholism, while others develop emotional, social or behavioral problems, given similar social history factors?"

This paper will review the research findings on causes of women's alcoholism and drug abuse, including: 1) correlates suggested by epidemiological studies, 2) social and psychological antecedents of abuse in women, 3) genetic and biological factors, and 4) recommendations for research. Because there is considerable room for speculation on this topic, the author has interpreted the existing information on the basis her own research and clinical work with drug-dependent women.

EVIDENCE FROM EPIDEMIOLOGY STUDIES

Percentages of Ever Using and Heavy Using Women. The percentage of women who have ever used alcohol appears to have increased from 1946 to 1968 (73% to 77%) (Cahalan, Cisin, & Crowley, 1969), remained constant until 1981 (Thompson & Wilsnack, 1984), and possibly decreased since 1981. Similar increasing and decreasing trends have been reported for high school (Johnston, et al., 1987) and college students (Engs and Hanson, 1988).

Although a smaller percentage of women than men drink heavily (6% vs 12%), there has been little change in these percentages of women who ever drink. Apparently, different etiological factors influence social drinking and problem drinking. Public opinion, prevention programs and legislation appear to have some impact on the percentage of women who choose to drink occasionally.. These prevention approaches, however, have little or no impact on heavy drinking or alcoholic women.

Who are the Heavy Users? Epidemiology surveys (Cahalan, et al., 1979; Clark & Midanick, 1982; Fillmore, 1984; Wechsler, 1980; Wilsnack, Wilsnack, & Klassen, 1981) consistently find that certain subpopulations of women are at high risk for heavy drinking—young adult women (21–24 years), middle-aged women (35–49 years), unmarried women (i.e., single, cohabiting, separated or divorced), and unemployed women. Middle to upper class women report higher heavy drinking rates (Cahalan, et al., 1969; Wechsler, 1980) and black women show a higher percentage of both heavy drinking and abstention

(Cahalan, et al., 1969). Seventy-one percent of all women with AIDS are Black or Hispanic, indicating high IV drug use in these minorities (Hopkins, 1987).

Epidemiological studies unfortunately report only correlations and incidence and prevalence rates, hence it is difficult to interpret reasons for these reported higher rates in certain subpopulations. This author believes dated surveys do not reflect rapid changes in social norms for heavy use in these subpopulations. Although middle and upper class women may have been "liberated" earlier to drink heavily like men, lower class and minority women have recently begun to catch up. Today, however, a new trend of concern with health and wellness which discourages heavy drinking is beginning in middle and upper class women.

These epidemiological surveys have found interesting correlations between subpopulations of heavy drinking women and certain social factors (i.e., role loss, unemployment, and alcohol-abusing parents, spouse, or friends) that are worth exploration in more controlled studies. One study found that 30 to 49 year old women who have experienced a role loss or 50 to 60 year old women who are unemployed and whose children have left home report higher drinking rates (Wilsnack, Wilsnack & Klassen, 1981).

Women abuse alcohol or drugs for the same reasons that men do—biological or genetic vulnerabilities, more life stressors than effective coping skills, depression, and a self-perception that condones self-medication. Because of the different roles men and women generally play in society, the specific reasons for women's unhappiness or stress often differ from men's. Also, because of the tremendous stigma placed on women who drink excessively or use illegal drugs, fewer women than men have a self-perception that includes abusing alcohol or illegal drugs. These same women often feel that it is acceptable to use prescription drugs and may self-medicate with a wide variety of medications.

PSYCHOLOGICAL AND SOCIAL ANTECEDENTS

Interaction of Environmental and Biological Factors. Environmental factors constantly interact with biological vulnerability factors. The author's Bio-psychosocial Model of Substance Abuse (Kumpfer, 1987) provides a complete framework of biological and environmental factors related to alcohol and drug abuse within a stress/coping/cognitions model. This framework helps answer Gomberg's question of why some women become substance abusers while others develop emotional and social problems under similar social stressors. The particular expression of a women's inability to cope with life stressors is heavily influenced by her biological makeup.

Women who develop alcoholism have a higher percentage of alcoholic relatives than women who develop other problems. Women most at risk are those with family history positive (FH+) for alcoholism on both sides of the family, but adoption studies suggest having an alcoholic biological mother is more salient than an alcoholic biological father (Bohman, Segvaardson, & Cloninger, 1981). An older Danish adoption study (Goodwin, 1977) found no higher rates of alcoholism in women with an alcoholic father, but higher rates of depression and an inability to make friends as children. While studies suggest that daughters of Type II (male-limited) alcoholics are not susceptible to this type of alcoholism, Bohman and associates (1984) found increased rates of Briquet's Syndrome (diversiform somatization). Clinicians are also reporting in non-controlled studies increases in eating disorders in these women, occasionally combined with alcoholism or drug abuse.

PSYCHOSOCIAL VULNERABILITIES

Women are very susceptible to social influences to abuse alcohol and drugs. Dysfunctional family environments in childhood and adulthood are often reported by women alcoholics. More women than men alcoholics report having an alcoholic parent, siblings or family members (Sherfy, 1955; Lisansky, 1958; Wood & Duffy, 1966; Winokur & Clayton, 1968). Likewise, more women than men alcoholics have alcoholic spouses (Sherfy, 1955; Lisansky, 1985).

Early Childhood Risk Factors. In a study of alcoholic women, Gomberg (1984) found a high percentage were children of alcoholics (COAs). They had poor relationships with their mothers, and they exhibited more emotional and behavioral problems as children (i.e., temper tantrums, phobias, loneliness, acting out, unpopularity, trouble in school and running away). Alcoholic women also report less approval from their parents and more feelings of childhood deprivation and lack of social support than non-alcoholic women (Corrigan, Schilit & Gomberg, 1987).

A disquieting picture of causes of chemical dependency in COA women is emerging for further verification. The scenario goes like this: A girl with genetic vulnerabilities is born to a dysfunctional family in which the mother tends to be cold and domineering and the father warm, gentle and often alcoholic (Kinsey, 1966; Wood & Duffy, 1966). The young girl rejects her mother and identifies with the father and his drinking. An incestuous relationship may develop as studies estimate that between 20 and 75% of adult women and 71 to 90% of teenage girls in alcohol treatment have been sexually abused (Gomberg & Lisansky, 1984; Rohsenow, et al., 1988). Sexual victimization has a direct effect on adolescent drug use whereas physical abuse has an indirect effect on drug use mediated by self-derogation (Dembo, et al., 1987). Role reversals can occur in which the child feels responsible for the parent's well-being and marriage. The young girl develops guilt and shame about the incestuous relationship. Often a love/hate relationship develops with the mother who allows this to happen.

In such a situation, a young woman is left with many mixed messages and feelings, a lack of basic trust in others, intense guilt and shame, lack of adequate coping skills combined with an immense feeling of responsibility for keeping the family together. Because of their stronger identification with their father, some COA women may turn to nontraditional sex roles, and occasionally lesbian relations. The rate of homosexuality is higher among chemically dependent females than other females, but causes of this are not known (Sandmaier, 1980). Sexual abuse and related family factors have been linked to drug abuse and delinquent behaviors (Burgess, Hartman, & McCormack, 1987).

Adolescent Risk Factors. Wilsnack and Thompson (1984) in a comparison of national surveys of adolescents identified those female adolescents at risk for heavy drinking as: 1) those with parents who drink; 2) those drinking at a later age; 3) those with more opportunities to drink; and 4) those not adhering to traditional general-specific norms and values. All of these factors are associated with COAs. Other correlates of alcohol abuse in women that could be related to being COAs is permissiveness of parents concerning drinking (Cahalan, et al., 1969; Thomas & Wilsnack, 1984). Parental approval or disapproval is an important factor in whether a female adolescent drinks (Rachal et al., 1982). If a strong bond is established with nonabusing parents, the adolescent female is less likely to be influenced by internalized peer norms to use. Lack of social support by parents is reported of alcoholic women (Schilit & Gomberg, 1987).

Early disruption or trauma are more common in women alcoholics than male alcoholics, such as divorce, death of a parent, or alcoholism (Curlee, 1970; de Lint, 1964; Lisansky 1957; Rosenbaum, 1958). Such stressors can lead even COAs or COSAs who vow: "It will never happen to me" (Black, 1983), into use when they do not have sufficient coping strategies and skills to handle these life stressors. High school COAs have higher rates of depression, low self-esteem and heavy drinking than their peers (Roosa, et al., 1988). COA females may be deficit in abstracting and problem-solving abilities (Turner & Parsons, 1988) to handle adolescent problems. In adolescence, loss of boyfriends, disappointments in school or extracurricular activities can lead biologically vulnerable COA women into abuse.

Adulthood Risk Factors. Depression is a common correlate of substance abuse in women, but whether depression is a primary cause of substance abuse or primarily a consequence may depend on the particular woman. Two types of alcoholism have been distinguished: 1) primary alcoholism and 2) secondary alcoholism (Schuckit, et al., 1969). Secondary alcoholism is preceded by depressive symptoms and is associated with an earlier age of onset of drinking and earlier loss of control (Turnbull & Gomberg, 1988). Hence, some young women may drink or use drugs to self-medicate undiagnosed clinical depressions of either the unipolar or bipolar affective type. Mello (1986) found that premenstrual dysphoria may increase alcohol and/or marijuana use at premenstruum. Adopted COAs have been found to be more depressed as adults and more friendless in children than adopted non-COAs (Goodwin, 1977). Women alcoholics have low self-esteem, poor self-concept, and feelings of inadequacy and futility. Such depressive symptoms may cause or be a consequence of chemical abuse.

Alcoholic women more often than males report a family crisis or traumatic event (e.g., divorce, gynecological problems, empty nest syndrome) preceding their drinking (Beckman, 1976; Curlee, 1970; Sandmaier, 1980; Wilsnack, 1973). Single parenthood combined with unemployment is a risk factor for substance abuse as is marriage combined with working. The stress of dual role demands can contribute to each substance abusers. Unfortunately, COA women often seek stressful care-giver occupations, such as nursing. Inability to cope with stress combined with a need for acceptance have been reported by Stammer (1988) as factors underlying substance abuse in COA nurses.

Women's choice of substance-abusing friends and spouses is associated with their abuse. Whether they choose such friends as a way to use or feel pressured to use is unclear. An assortive mating theory has been proposed by Rimmer and Winokur (1972), suggesting alcoholics choose to marry other alcoholics because of their similar backgrounds. However, older women alcoholics often report their alliance with an alcoholic as the primary cause of their abuse (Vaglum & Vaglum, 1987). Women alcoholics report their existing relationships as less happy and less supportive (Schilit & Gomberg, 1987) and divorce and separation rates are high (Wolin, 1980). Husbands are less tolerant than wives of a drinking spouse and divorce their spouse more often (Fox, et al., 1972).

BIOLOGICAL VULNERABILITIES

Genetic and Biological Factors. Since biblical times, many people have believed that "the sins of the fathers are visited upon the children." Plutarch, an early Greek writer, concluded that "drunkards beget drunkards." For many centuries people have believed that alcoholism runs in families and is inherited in a Lamarckian manner (i.e., if the father studied art, the children might inherit artistic talents; if the mother drank, the children might be drunkards). Of course, if the mother drinks, the children may develop developmental problems that would make them more prone to become alcoholics.

Few researchers today accept that chemical dependency is totally caused by genetic factors, but by a complex interaction of genetic, in utero developmental, and environmental factors. If chemical dependency is to be thought of as a "disease," it is mostly in the sense of a "disease of lifestyle" such as diabetes and heart disease, where genetic vulnerabilities can be triggered by stressors, nonsupportive lifestyles, and negative environments.

The field of alcoholism and recently drug abuse has turned attention towards research attempting to determine what exactly is being inherited by these offspring of chemically dependent parents that makes them more susceptible to alcohol and drug abuse later. Identification of biological vulnerabilities and markers would provide warning signals to affected persons that they should limit alcohol consumption and be sensitive to early signs of dependency. Once these precursors are understood, risk assessments could be developed and referrals made for prevention interventions designed to reduce each risk factor in a particular youth.

Biomedical research in this area is still in its infancy, and the few existing studies need additional replication; but a consistent picture is beginning to emerge of 1) differences in metabolism and reaction to alcohol and other drugs, 2) predisposing temperament and psychological characteristics, 3) neurological and biochemical differences, and 4) psychological and cognitive differences that could make a person more vulnerable to substance abuse (Kumpfer, 1987).

However, much of this research has been accomplished with males.

RECOMMENDATIONS FOR RESEARCH

There are several reasons for the lack of gender-specific research on women. Mello (1986) discussed the "Adam's Rib Syndrome," researchers tendency to behave as if women are identical men, as a deterrent to including women in research. From this perspective, studies on men should generalize to women and there is no reason to complicate the research with women. Another reason, mentioned by Hamilton (1986), is the preference for "basic" as opposed to "applied" research in advancing scientific knowledge. The basic mechanisms of behavior, learning, inheritance, or pharmacologic drug action are assumed to be essentially similar for both men and women. Another is the concern that hormonal cycles in women complicate any type of research with women, hence they are "too messy and complicated" to deal with.

The pressing need to deal with the increasing drug problem in women will hopefully force scientists to overcome their resistances to conducting research that will increase our knowledge of the etiology of drug abuse in women. Biological research contributing to etiological factors, including genetic twin and adoption studies as well as clinical pharmacology, neuropsychology, and neurochemistry studies, should include gender as a major variable. In psychosocial etiology studies should consider both proximal and distal antecedent in women. More complex theoretical models should be tested in longitudinal and cross-sectional studies, with separate modeling analyses for men and women. Promising leads from the existing research reviewed in this paper should be followed until confirmed or discarded. Federal research agencies should encourage principal investigators to include gender as a variable in their studies.

SUMMARY

This review of the research on the etiology of alcohol and drug abuse suggests that a large number of biological and psychosocial factors interact in complex ways to determine whether a particular woman will become an alcohol or drug abuser. While total accuracy in predicting human behavior may never be possible because of "sensitive dependencies on initial conditions" as specified by chaos theory, increased research on the etiology of substance abuse will help us design more effective prevention and treatment programs for women.

REFERENCES

Beckman, L. J. (1976). Alcoholism problems and women: An overview. In Greenblatt, M. and Schuckit, M. A. (eds.), *Alcoholism problems in women and children.* New York: Grune and Stratton.

Bohman, M., Sigvardsson, S. and Cloninger, R. (1981). Maternal inheritance of alcohol abuse: Cross-fostering analysis of adopted women. *Archives of General Psychiatry,* 38, 965–969.

Bull-Narc. (January–March, 1987). *Review of drug abuse and measures to reduce the illicit demand for drugs by region.* Division of Narcotic Drugs of the United Nations Secretariat, 39(1):3–30.

Burgess, A. W., Hartman, C. R., and McCormack, A. (November, 1987). Abused to abuser: antecedents of socially deviant behaviors. *American Journal of Psychiatry,* 144(11): 1431–6.

Calalan, D., Cisin, I. H. and Crossley, H. M. (1968). *American drinking practices: A national study of drinking behavior and attitudes* (Monographs of the Rutgers Center of Alcohol Studies, No. 6). New Brunswick, NJ: Rutgers Center of Alcohol Studies.

Clark, B., and Midanik, L., (1982). Alcohol use and alcohol problems among U.S. adults: Results of the 1979 national survey. In U.S. Department of Health and Human Services, *alcohol and health monograph No. 1: Alcohol consumption and related problems* (DHHS Publication No. ADM 82-1190). Washington, DC: U.S. Government Printing Office.

Corrigan, E. M. (1980). *Alcoholic women in treatment.* New York: Oxford University press.

Curlee, J. A. (1970). A comparison of male and female patients at an alcoholism treatment center. *Journal of Psychology,* 74, 239–247.

deLint, J. E. (1964). Alcoholism, Birth rank and parental deprivation. *American Journal of Psychiatry,* 120, 1062–1065.

Dembo, R., Dertke, M., La-Voie, L, Borders, S., Washburn, M., and Schmeidler, J. (March, 1987). Physical abuse, sexual victimization and illicit drug use: a structural analysis among high risk adolescents. *Journal of Adolescence,* 10(1): 13–34.

Engs, R. C. and Hanson, D. J. (November–December, 1988). University Students' Drinking Patterns and Problems: Examining the Effects of Raising the Purchase Age. *Public Health Reports,* 103(6): 667–673.

Fillmore, K. M. (1984). "When angels fall": Women's drinking as cultural preoccupation as a reality. In Wilsnack, S. C. and Beckman, L. J. (eds.) *Alcohol problems in women* (pp. 7–36). New York: Guilford Press.

Gomberg, E. S. (1974). Women and alcoholism. In Franks, V. and Burtle, V. (eds.), *Women in therapy.* New York: Brunner/Mazel.

Gomberg, E. S. (1976). Alcoholism in women. In Kissin, B. C. and Begleiter, H. (eds.), *The biology of alcoholism* (Vol. 4: Social aspects of alcoholism) pp.117–165. New York: Plenum Press.

Gomberg, E. S. (1980). Risk factors related to alcohol problems among women: Proneness and vulnerability. In *Alcoholism and alcohol abuse among women: Research Issues* (NIAAA Research Monograph No. 1, U.S. Department of Health, Education, and Welfare Publication No. ADM-80-835). Washington, DC: U. S. Government Printing Office.

Gomberg, E. S. L. and Lisansky, J. M. (1984). Antecedents of alcohol problems in women, In Wilsnack, S. C. and Beckman, L. J. (eds.), *Alcohol problems in women* (pp. 233–259). New York: Guilford Press.

Goodwin, D. W. (1987). Genetic influences in alcoholism. *Advanced Internal Medicine,* 32:283–97.

Goodwin, D. W., Schulsinger, F., Knop, J., Mednick, S., Guze, S. B. (1977a). Alcoholism and depression in adopted-out daughters of alcoholics. *Archives of General Psychiatry,* 34, 751–755.

Gorman, D. M. (October,1986). Comments on D. J. Cooke and C. A. Allen's Stressful Life events and alcohol abuse in women. *British Journal of Addiction,* 21(11): 1145–55.

Holder, H. D., Blose, J. O. (March, 1987). Reduction of community alcohol problems: computer simulation experiments in three counties. *Journal of the Studies of Alcoholism,* 48(2): 124–35.

Hopkins, D. R. (November–December, 1987). AIDS in minority populations in the United States. *Public Health Report,* 102(6): 677–81.

Kinsey, B. A. (1966). *The female alcoholic: A social psychological study.* Springfield, IL: Charles C. Thomas.

Kumpfer, K. (1987). Special populations: Etiology and Prevention of Vulnerability to chemical dependency in children of substance abusers in: Brown, B. S. and Mills, A. R. *Youth at Risk for Substance Abuse.* NIDA: Rockville, MD.

Lisansky, E. S. (1957). Alcoholism in women: social and psychological concomitants. I Social History data. *Quarterly Journal of Studies on Alcohol,* 18, 588–623.

Lisansky, E. S. (1958). The woman alcoholic. *Annals of the American Academy of Political and Social Sciences,* 315, 73–81.

Rachal, J. V., Williams, J. R., Brehm, M. L., Cavanaugh, B., Moore, R. P. and Eckerman, W. C. *Adolescent drinking behavior, attitudes and correlates: A national study* (Final report, Research Triangle Institute, RTI Project No. 23U-891). Research Triangle Park, NC: Research Triangle Institute.

Rachal, J. V., Maisto, S. A., Guess, L. L. and Hubbard, R. L. Alcohol use among youth. In *Alcohol consumption and related problems* (NIAAA Alcohol and Health Monograph No. 1, Pub. No. [ADM] 82-1190). Washington, DC: U. S. Government Printing Office.

Rohsenow, D. J., Corbett, R. and Devine, D. (1988). Molested as children: A hidden contribution to substance abuse? *Journal of Substance Abuse Treatment*, 5(1): 13–18.

Rosenbaum, B. (1958). Married women alcoholics at the Washingtonian Hospital. *Quarterly Journal of Studies on Alcohol*, 19, 79–89.

Sandmaier, M. (1980). *The invisible alcoholics: Women and alcohol abuse in America*. New York: McGraw-Hill.

Sherfey, J. M. (1955). Psychopathology and character structure in chronic alcoholism. In Diethelm, O. (ed.), *Etiology of chronic alcoholism* (pp. 16–42). Springfield, IL: Thomas.

Schilit, R. and Gomberg, E. L. (Summer, 1987). Social support structures of women in treatment for alcoholism. *Health-Society-Work*, 12(3):187–95.

Schuckit, M. A., Pitts, F. M., Jr., Reich, R., King, L. J. and Winokur, G. (1969). Alcoholism. I. Two types of alcoholism in women. *Archives of General Psychiatry*, 20, 301–306.

Stammer, M. E. (March, 1988). Understanding alcoholism and drug dependency in nurses. *Quarterly Review Bulletin*, 14(3): 75–80.

Thompson, K. M. and Wilsnack, R. W. (1984). Drinking and drinking problems among female adolescents: Patterns and influences. In Wilsnack, S. C. and Beckman, L. J. (eds.), *Alcohol problems in women* (pp. 37–65). New York: Guilford Press.

Turnbull, J. E. and Gomberg, E. S. (June, 1988). Impact of depressive symptomatology on alcohol problems in women. *Alcoholism in New York*, 12(3): 374–81.

Turner, J. and Parsons, O. A. (May, 1988). Verbal and nonverbal abstracting—problem-solving abilities and familial alcoholism in female alcoholics. *Journal of Studies in Alcoholism*, 49(3): 281–7.

Vaglum, S. and Vaglum, P. (November, 1987). Partner relations and the development of alcoholism in the female psychiatric patients. *Acta. Psychiatr. Scand.* 76(5): 499–506.

Wechsler, H. (1980). Epidemiology of male/female drinking over the last half century. In *Alcoholism and alcohol abuse among women: Research issues* (pp. 1–31). (NIAAA Research Monograph No. 1, U. S. Department of Health, Education and Welfare Publication No. ADM-80-835). Washington, DC: U.S. Government Printing Office.

Wilsnack, S. C. (1973). The needs of the female drinker: Dependency, power, or what? In *Proceedings of the Second Annual Alcoholism Conference of the National Institute on Alcohol Abuse and Alcoholism* (U. S. Department of Health, Education and Welfare Publication No. NIH-74-676). Washington, DC: U. S. Government Printing Office.

Wilsnack, S. C. and Beckman, L. J. (1984). *Alcohol problems in women*. New York: Guilford Press.

Wilsnack, S. C., Wilsnack, R. W., and Klassen, A. D. (1984/85). Sex differences and alcoholism in primary affective illness. *British Journal of Psychiatry*, 113, 972–979.

Wood. H. P. and Duffy, E. L. (1966). Psychosocial factors in alcoholic women. *American Journal of Psychiatry*, 123, 341–345.

Part C

PHYSIOLOGICAL ASPECTS

CHAPTER 4

Alcohol and Hormones: Reproductive and Postmenopausal Years[1]

Judith S. Gavaler, Ph.D.

INTRODUCTION

The effects of alcoholic beverage consumption on hormone levels and reproductive function in women have been described in a variety of settings. Although the effects themselves are generally known, the specific mechanisms involved have not yet been well elucidated. It has been suggested that sex bias on the part of investigators is responsible for the fact that women have been less well studied than men (Aron et al., 1965). A more likely explanation for our relative lack of knowledge in the particular area of alcohol effects on female hormones and reproductive function lies in the fact that it is difficult to study women because of the continual changes in hormonal status due to the menstrual cycle. With respect to the hormonal status of postmenopausal women, few studies have been reported, even though in postmenopausal women cyclic ovarian function has ceased and the complexity of hormonal variation is no longer a complicating factor in study design. That few studies are available concerning alcohol effects on postmenopausal hormonal status may reflect the fact that sustained interest in alcohol effects in general among the elderly is of fairly recent vintage (Blake, 1978).

Most studies available have evaluated alcoholic women and/or alcoholic women with alcohol-induced liver disease; as a result, most of the available data have been obtained in women who have abused alcoholic beverages for a prolonged period of time. Essentially no data are currently available from studies in women who use alcoholic beverages moderately during the reproductive years. Thus, while information is available concerning the effects on the menstrual cycle and hormone levels at the high dose end of the spectrum of alcohol use, there are no studies which document at what level of alcoholic beverage consump-

---

1. Project partially supported by NIAAA

tion effects begin to become evident. The situation in postmenopausal women is somewhat different in that, in addition to reports evaluating alcoholic postmenopausal women, recent studies have evaluated the effects on hormones of moderate consumption of alcoholic beverages by healthy postmenopausal women (Bo et al., 1982; Chaudhury & Matthews, 1966; Cranston, 1958).

In addition to studies in humans, considerable data obtained in experimental animals have increased our knowledge of alcohol effects not only on reproductive function, but also on hormone levels. Such studies have been performed both in animals with reproductive function and in animals which have had their ovaries surgically removed (ovariectomized), and thus are a model for some postmenopausal women. The animal studies have confirmed and extended the information obtained from studies in humans, and therefore animal research will continue to provide a powerful tool for elucidating the effects of alcohol and their mechanisms on the hormonal status of women in the future.

THE REPRODUCTIVE YEARS

The Menstrual Cycle

Sex hormones are produced and secreted by both the ovaries and the adrenal glands. During the reproductive years, the ovary is the primary source of the female sex hormones, estradiol and estrone (estrogens). The ovary is comprised of reproductive cells and endocrine cells. The reproductive cells, or germ cells, develop into a mature ovum (egg cell) during the menstrual cycle; at the time of ovulation, the follicle containing the ovum ruptures, releasing the egg, and the ruptured follicle then develops into the corpus luteum. The endocrine cells include granulosa cells which are responsible for estrogen production, the thecal cells which produce androgens that can be converted into estrogens, stromal cells which also produce androgens, and the corpus luteum which is responsible for progesterone production. Each of these separate components of the endocrine compartment of the ovary functions continuously but at quite different rates during the phases of the menstrual cycle. The granulosa cells are essential for estrogen synthesis and thus are required for follicular maturation, and the maintenance of female sexual characteristics. Thecal cells by augmenting estrogen production via the conversion of secreted androgens to estrogens help to provide the signal for the midcycle burst of gonadotropin secretion which is essential for ovulation. The corpus luteum by producing progesterone prepares the endometrium of the uterus for the fertilized egg and is essential for the maintenance of pregnancy until the placenta is capable of doing so.

The first part of the menstrual cycle is known as the follicular phase because the follicle with its contained egg cell is maturing during this time; levels of progesterone are low, while estrogen concentrations are first low and then rise during this part of the cycle. After ovulation, the second part of the menstrual cycle is known as the luteal phase because the corpus luteum, formed from the ruptured follicle, is producing progesterone during this time; levels of both estrogen and progesterone rise sharply and then fall during this phase of the cycle. The luteal phase is followed by the period of menstrual bleeding (menses) which occurs as progesterone production has diminished.

The ovary is part of the hypothalamic pituitary ovarian axis. The hypothalamus and pituitary gland are located at the base of the brain and are involved in the regulation of the menstrual cycle. The hypothalamus produces gonadotropin releasing factors which signal the pituitary gland to produce and secrete the gonadotropins luteinizing hormone (LH) and follicle stimulating hormone (FSH). FSH levels rise late in the preceding cycle and early in the next cycle, while LH rises just before ovulation and increases dramatically at midcycle (ovulation).

Expected Effects of Alcohol on the Menstrual Cycle

If alcohol were to affect the menstrual cycle, what effects of alcohol might be expected to occur? Overall, if alcohol interferes with reproductive function, then it would be expected that menstrual cycles might become irregular or that menses might not occur at all; the latter condition is called amenorrhea, and sustained irreversible amenorrhea prior to age 55 is known as early menopause. In both cases, alcohol might interfere with the development and maturation of the follicle with its contained egg cell. Similarly, alcohol might affect the endocrine cells and result in changes in the amounts of estrogens and progesterone produced. Additionally or alternatively, alcohol might affect the function of the hypothalamus and/or pituitary gland and cause changes in levels of the gonadotropins, LH and FSH. Any or all of these effects could be postulated to occur.

Alcohol Effects on Reproductive Function, the Menstrual Cycle and Hormone Levels

Chronic abuse of alcoholic beverages during the reproductive years has been reported to be associated not only with irregular menstrual cycles but also with early menopause (Table 1) (Eskay et al., 1981; Gaveler & Rosenblum, 1985; Gaveler, 1988; Gaveler et al., 1980; Gaveler, 1983). Studies in experimental animals have greatly extended these observations (Gaveler et al., 1987; Hugues et al., 1980; Hugues et al., 1978; James et al., 1982; Jones-Saunty et al., 1981; Kieffer & Ketchel, 1979;

McNamee et al., 1979; Mello et al., 1983; Mendelson et al., 1985; Mendelson et al., 1983; Mendelson et al., 1981; Merari, et al., 1973; Moskovic, 1975). In general, as the alcohol exposure either increases in dose or is prolonged over time, there is an increase in the number of female animals demonstrating a total loss of estrus cyclicity or a disruption of the existing cycles, usually with irregular prolongation of the cycles (Gaveler et al., 1987; Hugues et al., 1980; Hugues et al., 1978; James et al., 1982; Jones-Saunty et al., 1981; Kieffer & Ketchel, 1979; McNamee et al., 1979; Mello et al., 1983; Mendelson et al., 1985). Similarly, in rats, rabbits, mice and monkeys, an increased incidence of ovulatory failure documented by a failure to conceive, ovarian atrophy, absence of expected ova in the fallopian tubes following the time of expected ovulation, absence of ovarina corpora lutea, a loss of the midcycle ovulatory gonadotropin surge, and a failure of plasma estradiol and progesterone levels to increase in the luteal (latter) half of the cycle have seen following chronic ethanol administration to female animals (Hugues et al., 1980; James et al., 1982; Jones-Saunty et al., 1981; Kieffer & Ketchel, 1979; Mello et al., 1983; Mendelson et al., 1985; Mendelson et al., 1983; Mendelson et al., 1981; Merari et al., 1973). Not unexpectedly, as a result of this increased prevalence of ovulatory failure with increasing alcohol exposure, the fertility of alcohol-exposed female animals is affected adversely (Hugues et al., 1978; Mendelson et al., 1983; Moskovic, 1975).

While the effects of chronic exposure to alcohol are reproducible and consistent within and between studies in humans and experimental animals, the results of studies of the effects of ethanol administered acutely to normal women or animals are less satisfying, if only because they are relatively less encompassing. Specifically, although acute administration of ethanol to normal women over a period of several hours during the follicular phase of the menstrual cycle is reported to have no effect on hormone levels (National Inst. on Alcohol Abuse and Alcoholism, 1988; Ryback, 1977; Valimaki et al., 1984), when administered during the luteal phase, levels of estradiol, testosterone and progesterone are reported to be increased while gonadotropin concentrations are reduced (Valimaki et al., 1983). Studies in animals partially reproduce these findings in that acute administration of ethanol to monkeys during any phase of the menstrual cycle has no effect on estradiol levels (Rybak, 1977), while ethanol administered to rats immediately after the start of the LH surge is reported to cause LH levels to fall rapidly (Mendelson et al., 1981).

In summary, while alcohol administered acutely may or may not affect sex hormone levels and may cause a decrease in LH concentrations, chronic exposure to high doses of ethanol or alcoholic beverages clearly disrupts the menstrual cycle, inhibits ovulation and thereby may adversely affect fertility. No data are currently available concerning the effects of alcohol on the menstrual cycle of women who consume alcoholic beverages moderately during the reproductive years.

THE POSTMENOPAUSAL YEARS

Postmenopausal Endocrine Function

In postmenopausal women, the reproductive (follicular) compartment of the ovary by definition has ceased to function. The endocrine compartment of the postmenopausal ovary no longer contains functioning estrogen secreting granulosa cells, androgen secreting thecal cells or progesterone producing corpus luteum cells; the stromal cells, however, continue to produce the androgens testosterone and androstenedione. Because the postmenopausal ovary no longer produces estrogens, levels of estrogens are very low. Estrogen levels and the factors which may increase estrogen levels in postmenopausal women are important because the risk osteoporosis and cardiovascular disease are reported to be inversely related to estrogen concentrations.

The major source of estrogen in postmenopausal women is the conversion of androgens to estrogens. Thus, the continued ovarian stromal production of androgens is important because the conversion of androgens determines circulating levels of estrogens. Further, the production of androgens by the adrenal glands takes on great importance. Androgens are converted to estrogens in fat tissue, as well as in muscle and skin, by the enzyme called aromatase. Aromatase activity is elevated in postmenopausal women as compared to younger women and, in studies of males, is reported to be increased by alcohol. In addition, studies in male alcoholics report alcohol to increase the adrenal production of sex hormones.

In postmenopausal women the pituitary produces and secretes sustained high levels of the gonadotropins LH and FSH; the episodic secretion of gonadotropins associated with ovulation has ceased to occur. Postmenopausal levels of LH and FSH are inversely related to blood concentrations of estrogens.

Expected Effects of Alcohol on Postmenopausal Hormone Levels

If alcohol were to affect hormone production in postmenopausal women, what effects of alcohol might be expected to occur? By alcohol-induced increases in adrenal production of hormones, levels of the androgens testosterone and androstenedione might be expected to be increased; alternatively, if androgens are increasingly converted to estrogens, then levels of testosterone and/or androstenedione might demonstrate no change, or might even be decreased. By alcohol-induced increased conversion of androgens to estrogens, levels of estradiol and estrone might be expected to be increased. If levels of estrogen were to be elevated, then concentrations of the gonadotropins LH and FSH might be expected to be decreased.

Alcohol Effects of Postmenopausal Hormone Levels

As stated earlier, few studies evaluating alcohol effects on postmenopausal hormonal status have been performed. Effects of chronic alcoholic beverage abuse in amenorrheic and postmenopausal women with alcohol-induced cirrhosis has been reported to increase levels of estradiol, estrone and androstenedione (Table 1) (Van Thiel et al., 1985; Van Thiel et al., 1978). In women with alcohol-induced amenorrhea and early menopause, increased concentrations of estradiol, decreased levels of testosterone, and reduced levels of gonadotropins have been reported (Table 1) (Cranston, 1958; Gaveler, 1988; Gaveler, et al., 1980). Chronic administration of ethanol to ovariectomized rats has been shown to increase estradiol after exposure for 10 weeks, but no such increase was seen following only four weeks of exposure (Van Thiel et al., 1977); in these same experimental animals, no effect of ethanol on LH levels was seen at either four or 10 weeks (Vannicelli & Nash, 1984).

Two studies in normal postmenopausal women are available. In the first study, ethanol administered acutely produced no effect on LH concentrations. In the second study, chronic moderate alcoholic beverage consumption in healthy postmenopausal women was shown to result in a dose dependent increase in estradiol levels (Cranston, 1958).

In summary, the few studies available suggest that both chronic use as well as prolonged abuse of alcoholic beverages by postmenopausal women, in whom estrogen levels are low, produce detectable increases in circulating levels of estrogens, and result in reductions in concentrations of gonadotropins, at least in alcoholic postmenopausal women. These findings concerning alcoholic-induced increases in estradiol levels have been reproduced in an experimental animal model.

REFERENCES

Aron, E., Glanzy, M., Combescot, C., Puisias, J., Demaret, J., Reymouoard-Brault, G., Igberg, C. (1965) L'alcool, est-il dans le vin l'element qui perturbe chez la rat, le cycle vaginal? *Bull Acad Natl Med,* 149, 112–120.

Blake, C., (1978). Paradoxical effects of drugs acting on the central nervous system on the preovulatory release of pituitary luteinizing hormone in proestrus rats. *J. Endocrinol,* 79, 319–326.

Bo, W., Krueger, W., Rudeen, K., Symmes, S. (1982). Ethanol induced alterations in the morphology and function of the rat ovary. *Anat Rec,* 202, 255–260.

Chaudhury, R., Matthews, M., (1966). Effect of alcohol on the fertility of female rabbits. *J. Endocrinol,* 34, 275–276.

Cranston, E. (1958). Effect of tranquilizers and other agents on sexual cycle of mice.*Proc Soc Exp Biol Med,* 98, 320–322.

Eskay, R., Ryback, R., Goldman, M., Majchrowicz, E., (1981). Effect of chronic ethanol administration on plasma levels of LH and the estrous cycle in the female rat. *Alcoholism: Clin Exp Res,* 204–206.

Gavaler, J., Rosenblum, E. (1985). Exposure dependent effects of ethanol administered in drinking water on serum estradiol and uterus mass in sexually mature oophorectomized rats: A rat model for bilaterally ovariectomized/postmenopausal women. *J. Stud Alc,* 48, 295–303.

Gavaler, J. S. (1988). Effects of moderate consumption of alcoholic beverages on endocrine function in postmenopausal women: Bases for hypoteses. *Recent Developments in Alcoholism,* Vol 6, Plenum Publishing Corp., New York.

Gavaler, J., Van Thiel, D., Lester, R., (1980). Ethanol: A gonadal toxin in the mature rat of both sexes. *Alcoholism: Clin Exp Res,* 4, 271–276.

Gavaler, J., (1983). Sex-related differences in ethanol-induced hypogonadism and sex steroid responsive tissue atrophy: analysis of the weaning ethanol fed rat model using epidemiologic methods. In: T. J. Cicero (Ed) *Ethanol Tolerance and Dependence: Endocrinologic Aspects.* Alcohol and Health Monograph Series, No. 13. National Institute on Alcohol Abuse and Alcoholism, DHHS Pub. No (ADM) 83-1258. U.S. Government Printing Office, Washington, D.C. 78–88.

Gavaler, J. S., Belle, S., Cauley, J. (1987). Effects of moderate alcoholic beverage consumption on estradiol levels in normal postmenopausal women. *Alcoholism: Clin Exp Res,* 11. 199.

Hugues, J., Coste, T., Perret, G., Jayle, M., Sebaoun, J., Modigliani, E. (1980). Hypothalamo-pituitary ovarian function in thirty one women with chronic alcoholism, *Clin Endocrinol,* 12, 543–551.

Hugues, J., Perret, G., Adessi, G., Coste, T., Modigliani, E. (1978). Effects of chronic alcoholism on the pituitary-gonadal function of women during menopause transition and in the postmenopausal period, *Biomedicine,* 29, 279–283.

James, V., Green, J., Walker, J., Goodall, A., Short, Fl, Jones, D., Noel, C., Reed, M., (1982). The endocrine status of postmenopausal cirrhotic women. In: M. Langer, L. Chandussi, I. Chopra, L. Martini (Eds.). *The Endocrines and the Liver,* Serono Symposium No. 51, Academic Press, New York, 417–419.

Jones-Saunty, D., Fabian, M., Parsons, O. (1981). Medical status and cognitive functioning in alcoholic women. *Alcoholism: Clin Exp Res,* 5, 372–377.

Kieffer, J., Ketchel, M. (1979). Blockade of ovulation in the rat by ethanol. *Acta Endocrinol,* 65, 117–124.

McNamee, B., Grant, J., ratcliffe, J., Ratcliffe, V., Oliver,J., (1979). Lack of effect of alcohol on pituitary gonadal hormones in women. *Br J Addict,* 74, 316–317.

Mello, N., Bree, M., Mendelson, J., Ellinghoe, J., (1983). Alcohol self-administration disrupts reproductive function in female Macaque monkeys. *Science,* 211, 677–679.

Mendelson, J., Mello, N., Ellingboe, J., Bavli, S. (1985). Alcohol effects on plasma luteinizing hormone levels in postmenopausal women. *Biochem Behav,* 22,233–236.

Mendelson, J., Mello, N., Bavli, S., Ellingboe, J., Bree, M., Harvey, K., King, N., Seghal, P. (1983). Alcohol effects on female reproductive hormones. In: T. J. Cicero (Ed) *Ethanol Tolerance and Dependence: Endocrinologic Aspects.* Alcohol and Health Monograph Series No. 13. National Institute on Alcohol Abuse and Alcoholism, DHHS Pub. No (ADM) 83-1258. U.S. Government Printing Office, Washington,D.C. 146–161.

Mendelson, J., Mello, N., Ellingboe, J., (1981). Acute alcohol intake and pituitary gonadal hormones in normal human females. *J Pharmacol Exp Ther,* 218, 23–26.

Merari, A., Ginton, A., Heifez, T., Lev-Ran,T., (1973). Effects of alcohol on the mating behavior of the female rat. *Quart J Stud Alc,* 34, 1095–1098.

Moskovic, S., (1975). Uticaj hronicnog trovanja alkoholom na ovarijumsku disfunkciju (Effect of chronic alcohol intoxication on ovarian dysfunction). *Srp Arh Celok Lek,* 103, 751–758.

National Institute on Alcohol Abuse and Alcoholism (1988). Alcohol and Aging. *Alcohol Alert,* 2, U.S. Department of Health and Human Services, Washington, D.C.

Ryback, R., (1977). Chronic alcohol consumption and menstruation. *J Am Med Ass,* 238, 2143.

Valimaki, M., Pelkonen, R., Salaspuro, M., Harkonen, M., Hirvonen, E., Ylikahri, R., (1984). Sex hormones in amenorrheic women with alcoholic liver disease. *J Clin Endocrinol Metab,* 59, 133–138.

Valimaki, H., Harkonen, M., Ylikahri, R. (1983). Acute effects of alcohol on female sex hormones. *Alcoholism: Clin Exp REs,* 7, 289–293.

Van Thiel, D., Rosenblum, E., Phol, C., Gavaler, J. (1985). Lack of an effect of ethanol administered in drinking water to sexually mature oophorectomized rats on LH levels. *Alcoholism: Clin Exp Res,* 9, 194.

Van Thiel, D., Gavaler, J., Lester, R. (1978). Alcohol-induced ovarian failure in the rat. *J Clin Invest,* 61, 624–632.

Van Thiel, D., Gavaler, J., Lester, R. (1977). Ethanol: A gonadal toxin in the female. *Drug Alc Depend,* 2, 373–380.

Vannicelli, M., Nash, L. (1984). Effect of sex bias on women's studies in alcoholism. *Alcoholism: Clin Exp Res,* 8, 334–336.

TABLE 4.1 Effects of Chronic Alcoholic Beverage Use and Abuse in Women.

Author (Year)	Findings
The Reproductive Years	
Moscovic (1975)	Among 321 alcoholic women, 61% experienced menstrual irregularity problems and 8.7% experienced early menopause.
Ryback (1977)	In one alcoholic woman, amenorrhea dissipated with alcohol abstinence; in another alcoholic woman, amenorrhea dissipated but then reoccurred, even with continued alcohol abstinence.
Hugues et al (1980)	Among 31 alcoholic women, menstrual irregularity problems and early menopause reported.
Jones-Saunty et al (1981)	Among 100 alcoholic women, menstrual irregularity problems and early menopause reported.
Valimaki et al (1984)	Among 9 amenorrheic alcoholic women, estrone reported to be increased, and estradiol, testosterone and progesterone, as well as LH and FSH, reported to be decreased.
The Postmenopausal Years	
Hugues et al (1978)	Among 11 alcoholic women, early menopause and decreased levels of LH reported.
Hugues et al (1980)	Among 21 alcoholic women, estradiol reported to be increased, and levels testosterone, LH and FSH reported to be reduced.
James et al (1982)	In 2 alcoholic women with alcohol-induced cirrhosis, levels of estrone, estradiol and androstenedione reported to be increased.
Gavaler et al (1987)	Among 129 healthy postmenopausal women, estradiol reported to be increased in a dose-dependent manner with moderate alcoholic beverage consumption.

CHAPTER 5

Physiological Effects of Cocaine, Heroin and Methadone

Janet L. Mitchell, M.D., M.P.H.
Gina Brown, M.D.

INTRODUCTION

The abuse of drugs has become widespread in American society. The ramifications for individual health is great and the effect on health care costs is overwhelming. Woman comprise an ever increasing number of heroin, cocaine and methadone users. Along with alcohol, barbiturate, and other prescription drugs, abuse of these drugs profoundly impacts the health status of women. While data on the relationship of drug addictive behavior to particular illnesses i.e. AIDS and other HIV related diseases is well known, very little is written about the physiologic effect of individual drugs on women. The literature that does address addiction in women is largely limited to the psychology of substance abuse or the impact of perinatal drug use on neonatal outcome.

Studies that directly address reproductive function in females are primarily based on studies done in laboratory animals. While some of the outcomes observed in animals have been seen clinically, the scope of the drug effect has been limited by our inability to conduct these potentially harmful studies on humans. The additional variables of poor nutrition and poly drug use further confuse the evaluation of drug abuse in humans. Each of the above mentioned drugs has profound physiologic effects in various animal models that can only be extrapolated to predict the potential impact on women.

COCAINE

Cocaine is obtained from the leaves of the ERYTHROXYLON COCA plant. In its acidic form, cocaine hydrochloride is a white powder used intranasally or in an intravenous solution. The alkaloid form of cocaine known as free-base cocaine, crack, or rock cocaine is heat stable and its use is by smoking. The intranasal route of cocaine inhibits the drug's inherent toxicity by its vasoconstrictive action which limits its absorption across mucous membranes. The alkaloid form provides increased drug availability over the extensive pulmonary vasculature. Its slow absorption across mucous membranes provides a prolonged and extensive effect and intensifies its toxicity. It is detoxified by hepatic and plasma cholinesterase and excreted via the kidney (Cregler, 1989).

Cocaine acts in the central and peripheral nervous systems, providing an excess of neurotransmitter at the synapse. Its peripheral mechanism of action is to prevent the re-uptake of norepinephrine at the synaptic level and to prolong catecholamine stimulation of the receptors. Centrally, it also acts to increase dopamine concentration at the receptor level. The increase in concentration of norepinephrine may cause a marked rise in blood pressure, vasoconstriction and tachycardia (Cregler, 1989; Cregler, 1988; Bates, 1980).

Enhanced beta stimulation of the myocardium has been known to cause cardiac arrhythmias. The vasoconstrictive effect of the drug has been one of the factors associated with cocaine related myocardial infarction. It is postulated that cocaine induced coronary artery vasospasm leads to myocardial hypoxia. Vasospasm coupled with the increased myocardial oxygen demand caused by cocaine induced tachycardia may result in myocardial infarction (Cregler, 1988).

In pregnancy, the vasospasm and hypertensive effect of cocaine has been associated with placental abruption (Acker, 1983; Chasnoff, 1985). A review of six cases reported in the literature presented with vaginal bleeding and contractions which began soon after cocaine usage. At delivery, the patients were found to have a large retro-placental clot indicative of an abruption. The patients also displayed transient hypertension which is consistent with recent cocaine use.

Seizures, one of the earliest identified side effects of cocaine use, may be the result of drug induced cardiac arrythmia or associated hyperpyrexia. Cocaine also has a direct effect of lowering the seizure threshold (Jonsson, 1983). At Harlem Hospital Center, the management of a number of cases thought to be eclampsia was complicated by the patients' recent ingestion of cocaine (unpublished data).

Central nervous system toxicity for cocaine is both direct and indirect. Cerebrovascular accidents occurring in temporal relationship to cocaine usage have been described in young adults and in a neonate (Cregler, 1989; Cregler, 1988; Chasnoff, 1985). The varied causes include aneurysms, arteriovenous malformations, and arterial occlusion. Chasnoff (1988) described a neonate who developed a cerebral infarct after birth. The infant exhibited decreased right sided tone, seizures, tachycardia and hypertension after birth. CAT-scan revealed an acute infarct in the distribution of the left middle cerebral artery. The infant's mother had used cocaine intranasally fifteen hours before delivery. Because the infant showed symptoms of weakness at birth, it was thought that the infarction occurred prenatally. In another case at Harlem Hospital, a pregnant patient with a previously documented, inoperable, arterio-venous malformation, and no residual symptomatology, exhibited transient symptoms of unilateral weakness temporally associated with her use of free-base cocaine.

The indirect, centrally mediated effects of cocaine are due to the neurotransmitter effects on hormone release. In ovariectomized female rats, cocaine in low to moderate doses was shown to lead to an increase in luteinizing hormone (LH) levels. High doses of cocaine inhibited LH release. Follicular stimulating hormone (FSH) was not affected by the drug. Prolaction was decreased by cocaine at low and high doses (Smith, 1987; Gawin, 1985).

In a study of men and women by Gawin and Kleber (Gawin, 1985), the cocaine associated decrease in prolactin levels seen in animals was substantiated. In a follow up study, chronic cocaine abusers undergoing rehabilitation (Cocores, 1986), exhibited hyper-prolactinemia and decreased libido in seven out of the ten patients studied. One patient developed galactorrhea. This alteration in LH and prolactin levels led to amenorrhea and oligomenorrhea. The effects were reversed by bromocriptine.

HEROIN

Morphine, the alkaloid form of opium, was first isolated in 1803 (Sternbach, 1980). The acetylation of morphine produces heroin which has two to four times the analgesic power of its parent drug. Originally thought to be a safe, non-addictive substitute for morphine, heroin has become one of the major drugs of abuse in the world. Used intravenously, subcutaneously and intranasally, heroin is rapidly converted to morphine. The greater lipid solubility of heroin in comparison to morphine results in a more rapid crossing of the blood brain barrier when administered intravenously. This causes a rapid production of the CNS effects. Once in the brain, heroin is rapidly de-acetylated to morphine and exerts its CNS agonist action as an analgesic (Jaffe, 1985).

The toxic effects of heroin are as much related to the method used for preparation as to the direct actions of the drug. While quinine is one of the more popular diluents for both users and pushers; sucrose, lactose, talc, procaine, mannitol and baking soda are also used. The bitter, narcotic-like taste of quinine makes it palatable to sellers of the drug. The flash sensation produced at injection is secondary to the dilation of facial vessels and makes this diluent a favorite among heroin abusers. Quinine toxicity which includes damage to the CNS auditory apparatus, skeletal, cardiac and smooth muscle, kidneys and gastrointestinal tract has been documented (Sternbach, 1980).

Intravenous and subcutaneous injection of heroin results in extensive skin ulcerations. 'Skin popping', more common among female addicts, results in abscesses and cellulitis (Jaffe, 1985). Other common sequelae of heroin addiction which are largely infectious, are due to the needle sharing habits of drug abusers and to their lack of sterile technique during injection. While bacterial endocarditis and septic thrombophlebitis are common causes of death among intravenous heroin users, the spread of HIV through needle sharing has markedly increased the morbidity and mortality among female abusers. More than half of the women reported to CDC with AIDS list their own IV drug use as their risk factor.

Like cocaine, heroin and other opiates have been shown to act centrally at the hypothalamic level to alter hormone secretion and to disrupt the menstrual cycle (Gawin, 1985). Narcotic induced reductions in LH and FSH levels have been observed in humans and female rats. Secondary amenorrhea is common in female heroin abusers. In a study of seventy-six former heroin addicts receiving daily methadone maintenance, more than half experienced menstrual abnormalities while using heroin or methadone (Santen, 1975). Studies of adolescent girls using narcotics showed that one-third to two-thirds stopped menstruating during their period of drug abuse. Oligomenorrhea and hypomenorrhea was also demonstrated in another group. These sequelae often persisted for as long as a year after the drug abuse was discontinued (Litt, 1975; Prather, 1978; Bai, 1974; Litt, 1970).

A survey of 100 women inpatient at the federal facility for the treatment of narcotic addiction located in Lexington, Kentucky provides additional information about the reproductive and gynecologic health of these women. Given the limitations of self-report with no substantiating laboratory data, of the 81 heroin abusers, 3% reported amenorrhea and 10%, oligomenorrhea during their period of abuse. The average length of amenorrhea was 17.3 months. With the discontinuation of heroin, menses returned after an average of 1.8 months. Although 63% thought themselves infertile at some point during their addiction, there were 77 pregnancies in 39 of the same women. Sixty-five per cent also reported dysmenorrhea severe enough to necessitate absence from work or school. Thirty-one per cent reported dysmenorrhea prior to drug use, 22% during their addiction and 60% after cessation of drugs. This increase after drug use was felt by the authors to be related to multiple factors. The high incidence of venereal disease in this population of women can lead to pelvic inflammatory disease. A higher threshold to pain during addiction and the infrequency of menses with heroin abuse may also be a contributing factor. The tendency to malinger in this group of women, especially after drug cessation and evidence that addicts may be more pain prone may also play a role. The possibility that there may be some unknown endocrine effect of drug abuse cannot be ruled out (Stoffer, 1968).

Other effects of narcotics on sexual function have been observed. A dichotomy exists between the side effects of initial and extended drug use. The initial intravenous injection of heroin produces a 'rush' described as a warm flush of the skin and lower abdominal sensations likened to sexual orgasm in intensity and quality (Santen, 1975). However, in a study rating the subjective feelings of the rush, pleasure, relaxation, warmth and thirst ranked highest, while sexual orgasm ranked only fifteenth out of the twenty feelings listed (Seecoff, 1986).

Sexual dysfunction in the intravenous drug abuser has been well documented. In women who chronically abuse high doses of heroin, lack of sexual desire and decreased performance are not uncommon (Smith, 1982). However, many of them may be anorgasmic without the use of drugs (Seecoff, 1986).

METHADONE

Methadone, synthesized by the Germans, came into clinical use at the end of World War II. Although a potent analgesic, today its most prominent use is in the chemotherapeutic treatment of heroin or other narcotic addiction. Although structurally, it only remotely resembles morphine, the pharmacological properties are qualitatively identical and thus has the same effects as heroin. The outstanding properties of methadone are its effective analgesic activity even when given orally, its extended duration of action, its use in suppressing withdrawal symptoms in the treatment of opiate addiction and its tendency to show persistent effects with repeated administration (Jaffe, 1985). While it is commonly used as chemoprophylaxis is heroin addiction it is also a drug of abuse.

A detailed study of the menstrual dysfunction in seven methadone maintained women with endocrine profiles was published in 1975 by Santen et al (Santen, 1975). These seven women were drawn from a larger group of women (76) who were surveyed about their menstrual histories while on street heroin and on methadone maintenance. Only thirty per cent of these women experienced regular menses while using heroin or methadone. Ninety-five per cent of these same women reported regular menses prior to their addiction. The menstrual abnormalities in the women on methadone did not appear to be dose related. Menstrual function seemed to improve with long-term methadone ingestion.

Four of the seven women with extensive hormonal studies showed abnormalities in cyclic gonadotropin release. Four women who were ovulatory during the study averaged follicular and luteal phase LH secretory spikes similar to those reported in normal women. The estrogen negative feedback mechanism was not abolished by methadone although testing of the positive feedback mechanism was variable and dose related. For those women who ovulated, ovarian steroidal function appeared normal as did uterine response. In summary the authors felt this group of women resembled other women with "hypothalamic amenorrhea" in that the specific abnormalities leading to menstrual dysfunction were varied.

Studies that have looked at thyroid function in pregnant and non-pregnant women maintained on methadone have found no difference compared with the drug free control group although animal models have shown some effect on metabolism (Jhaveri, 1980).

A 1973 report on urine slide pregnancy tests showed an increased false positive rate on urine from women on various dosages of methadone. This rate seemed to increase with increasing dosages over 100 mg/day. The authors admitted, however, that this false positive rate might be related to the proteinuria seen in patients on very high dosages of methadone or to other drugs that might be present in the urine (Horwitz, 1973).

Of particular interest for those caring for pregnant women is the report of altered pharmacokinetics in methadone maintained pregnant women. In this report, Pond et al (1985) studied the effects of pregnancy on the maternal disposition of methadone in nine reportedly healthy women maintained on methadone for at least two months. While animal studies suggested that methadone metabolism was impaired during pregnancy, these authors found the opposite in women. The average duration of methadone maintenance in these nine women was 1.4 years. The women were studied at 27 and 37 weeks of gestation and at 2 and 8 weeks post partum. Though plasma concentrations taken immediately before the morning dose of methadone were significantly lower during pregnancy than in the post partum period. Total (bound plus unbound) and unbound plasma methadone clearance rates were higher during pregnancy than post partum. This suggests that some methadone maintained women may need an increase in dosage during pregnancy. It also supports the clinical observation that some women report waking up in the middle of the night with signs and systems of withdrawal. A twice-a-day dosage may benefit these women.

The question of whether women can be safely detoxified from methadone during pregnancy is still controversial. While there are reports of many successes, none have been documented in the literature.

Zuspan's article of 1978 (Zuspan, 1978) suggesting detoxification may create fetal distress leading to fetal demise is still unchallenged.

The question as to whether women on methadone should breast feed was also addressed in the Pond article (1985). The ratio of breast milk concentrations to plasma in two patients was extremely low. Based on these two patients, they calculated that infants would ingest between 0.01 and 0.03 mg of methadone/day based on an average daily consumption of 450 ml of breast milk/day.

SUMMARY

What is evident in exploring the effects of these drugs on women is that very little research has been done. Most studies center on the effects on the fetus and on the menstrual cycle. Little is known about the direct effects on fertility, menopause or menarche (a concern given the younger ages of addiction), or of the interaction of estrogen or progesterone with these drugs. Many articles that looked at a hormonal profile of methadone maintained patients that included prolactin included only men or only one or two women. The numbers of women presenting with complication of cocaine abuse during pregnancy and HIV related disease secondary to their own IV drug use are only a fraction of women engaged in these practices. It is important that research efforts be targeted toward women in order to optimize their treatment.

REFERENCES

Acker, D., Sachs, B. P., Tracey, K. J., Wise, W. E., (1983). Abruptio placentae associated with cocaine use. *Am J Obstet Gynecol*, 146, 220–221.

Bai, J., Greenwald, E., (1974). Drug related menstrual aberrations. *Obstet Gynecol*, 44, 713–719.

Bates, C. K., (1980). Medical risks of cocaine use. *West J Med*, 148, 440–444.

Chasnoff, I. J., Burns, W. J., Schnoll, S. H., Burns, K. A. (1985). Cocaine use in pregnancy, *NEJM*, 312, 666–669.

Cocores, J. A., Dackes, C. A., Gold, M. S., (1986). Sexual dysfunction secondary to cocaine abuse in two patients. *J Clin Psych*, 147, 569.

Cregler, L. L., (1989). Adverse health consequences of cocaine abuse. *J of Natl Med Ass*, 81, 27–38.

Cregler, L. L., Mark, H., (1988). Special Report: Medical complication of cocaine abuse. *NEJM*, 315, 1495–1500.

Gawin, F. H., Kleber, H. D., (1985). Neuroendocrine findings in chronic cocaine abusers: a preliminary report. *Br J Psych*, 147, 569.

Horwitz, C. A., Maslansky, R., Waldinger, R., Cabrera, H., Ward, P. C. J., (1973). Effects of methadone on human pregnancy tests. *J Reprod Fert*, 33, 489–493.

Jaffe, J., (1985). Drug addiction and drug abuse. In Goodman and Gilman's the Pharmacological basis of therapeutics, 7th ed., edited by A. G. Gilman, L. S. Goodman, T. W. Rall and T. Murad. MacMillian Press, New York, NY, 532–581.

Jaffe, J., Martin, W. (1985). Opioid analgesics and antagonists. In Goodman and Gilman's the Pharmacological basis of therapeutics, 7th ed., edited by A. G. Gilman, L. S. Goodman, T. W. Rall and T. Murad. MacMillian Press, New York, NY, 491–531.

Jhaveri, R. C., Glass, L., Evans, H. E., Dube, S. D., Rosenfeld, W., Khan, F., Salazar, J. D., Chandavasu, O. (1980). Effects of methadone on thyroid function in mother, fetus and newborn. *Pediatrics,* 65, 557–561.

Jonsson, S., O'Mera, M., Young, J. B. (1983). Acute cocaine poisoning: importance of treating seizures and acidosis. *Am J Med,* 75, 1061–1064.

Litt, I. F., Schonberg, S. K., (1975). Medical complications of drug abuse in adolescents. *Med Clin N Am,* 59, 1445–1452.

Litt, I. F., Cohen, M. I., (1970). The drug using adolescent as a pediatric patient. *J Pediatrics,* 77, 195–202.

Pond, S. M., Kreek, M. J., Tong, T. G., Raghunath, J., Benowitz, N. L. (1985). Altered methadone pharmacokinetics in methadone-maintained pregnant women. *J Pharmacol Exp Ther,* 233, 1–6.

Prather, J. E., Fidell, L. S., (1978). Drug use and abuse among women: an overview. *Int J Addict,* 13, 863–885.

Santen, F. S., Sofsky, J., Bilic, N., Lippert, R., (1975). Mechanism of action of narcotics in the production of menstrual dysfunction in women. *Fertility and Sterility,* 26, 538–548.

Seecoff, R., Tennant, F. S., (1986). Subjective perceptions to the intravenous "rush" of heroin and cocaine in opioid addicts. *Am J Drug and Alcohol Abuse,* 12, 79–87.

Smith, D. E., Moser, C., Wesson, D. R., Apter, M., Buxton, M. E., Davison, J. V., Orgel, M., Buffum, J., (1982). A clinical guide to the diagnosis and treatment of heroin related sexual dysfunction. *J of Psychoactive Drugs,* 14, 91–99.

Smith, C. G., Asch, R., (1987). Drug abuse and reproduction. *Fertility and Sterility,* 48, 355–373.

Sternbach, G., Moran, J., Eliastram, M., (1980). Heroin addiction: acute presentation of medical complications. *Ann Emerg Med,* 9, 161–169.

Stoffer, S., (1968). A gynecologic study of drug addicts. *Am J. Obstet Gynecol,* 101, 779–783.

Zuspan, F. P., (1978). Maternal intrauterine amine alterations in the pregnant drug addict. *J Reprod Med,* 20, 329–332.

CHAPTER 6

Alcohol, Pregnancy, and Fetal Development

Lyn Weiner, M.P.H.
Barbara A. Morse, Ph.D.

INTRODUCTION

Ethanol and its metabolites have the potential to cause a wide range of physiologic and neurologic disturbances in fetal development (Rosett and Weiner, 1984). The most extreme manifestation of alcohol's effects, fetal alcohol syndrome (FAS), includes growth retardation, facial dysmorphology, central nervous system anomalies and morphologic abnormalities. Less seriously affected people, who demonstrate morphologic and/or neurologic anomalies in the absence of the syndrome, are considered to have "possible fetal alcohol effects" (FAE).

Incidence figures for FAS are estimated to be 1.9 per thousand live births (Abel and Sokol, 1987). It is likely that FAE occurs three times more frequently than FAS. The incidence rate is difficult to specify as the symptoms are non-specific.

All children with FAS have been born to women who drink abusively; however every pregnancy which is complicated by alcohol abuse does not result in adverse effects. Prospective studies demonstrated FAS in 2.5% to 10% of the infants born to heavy drinkers (Sokol, 1981; Rosett et al, 1983b). The clinical and experimental literature provides an ever-increasing understanding of why the risks are not equal.

Alcohol's effects can result from both direct and indirect actions on the maternal-placental-fetal system throughout the gestational period (Rosett and Weiner, 1984). Biochemical and pathophysiologic effects of ethanol and its first metabolite, acetaldehyde, can disturb fetal development by disrupting cell differentiation and growth. High blood alcohol concentrations can induce alterations in maternal physiology and the intermediate metabolism of carbohydrates, proteins and fats which can alter the environment in which the fetus develops. Alcohol can interfere with the passage of amino acids across the placenta and with the incorporation of amino acids into proteins (Fisher et al, 1982). The variability in the nature and extent of abnormalities is related to several factors, including dose, chronicity of alcohol use, gestational stage and duration of exposure, and sensitivity of fetal tissue.

MEDIATING FACTORS

Experimental models, in which precise doses are administered and monitored, have demonstrated a clear dose-response effect on birthweight, mortality, and morphologic anomalies (Chernoff, 1977; Randall and Taylor, 1979). Peak blood ethanol concentration rather than alcohol dose has been demonstrated to be the critical factor in reducing brain growth in the rat (Pierce and West, 1986). In addition to a dose response effect, a threshold for alcohol's effects has been demonstrated in the rodent (Anders and Persaud, 1980; Skosyreva, 1973; Samson and Grant, 1984).

In prospective and retrospective studies, the relationship between heavy drinking and intrauterine growth retardation has been observed repeatedly (Rosett and Weiner, 1984). The severity and frequency of cranio-facial, morphologic, and skeletal abnormalities increased with dose (Ernhart et al, 1987). Risk was greatest with consumption greater than 3 ounces of absolute alcohol a day.

The dose of alcohol to which the fetus is exposed is directly related to maternal blood alcohol concentration (BAC). Alcohol reaches the fetus through the process of diffusion, crossing the placental membranes easily in both directions at a rate dependent on the concentration gradient. Following ingestion, the maternal BAC rises rapidly. Alcohol diffuses from the maternal system to the fetal system until equilibrium is reached; then the alcohol diffuses from the fetal circulation back into the maternal circulation (Pratt, 1980). Since alcohol dehydrogenase in the fetal liver is immature and has limited capacity, the fetus is dependent on the maternal system to eliminate the alcohol. Maximum concentrations occur later in the fetus than in the mother. Although the fetal BAC peaks at a lower point than that of the mother, the fetal BAC is slightly higher than the maternal BAC as they both approach zero. When the mother drinks continuously and maintains a relatively steady BAC, the difference between the maternal and fetal BAC is small.

Clinically, outcome has been associated not only with the amount of alcohol consumed, but also with the chronicity of maternal alcoholism. The severity of FAS has been more strongly associated with the stage of maternal alcoholism than with the absolute amount of alcohol consumed (Abel and Sokol, 1987; Majewski, 1981).

Gestational stage

Vulnerability of particular organ systems may be greatest at the time of their most rapid cell division. During the first trimester, effects of high concentrations of alcohol on cell membrane and cell migration can disturb embryonic organization of tissue, resulting in morphologic abnormalities. Throughout pregnancy, disturbances in the metabolism of carbohydrates, lipids, and proteins, and synthesis of RNA and DNA can retard cell growth and division. The third trimester is the time of rapid brain growth and neurophysiologic organization. High blood alcohol concentrations during this period may impair central nervous system growth and development and limit future intellectual and behavioral capacities.

The role of the timing of exposure has been clearly demonstrated experimentally. Rats and mice exposed to high doses of ethanol in week three (late gestation) experienced growth retardation similar to that caused by alcohol exposure throughout pregnancy (Lochry et al, 1982). When alcohol exposure was limited to early pregnancy, there was no significant reduction in birth weight. Abnormalities occurred in the developing brain of the mouse when high doses of alcohol (350–500 mg/100 ml) were administered on day 7 (Webster et al, 1980). High ethanol concentrations on day 8 were associated with maxillary hypoplasia and on days 9, 10, and 11 with skeletal anomalies.

Genetic susceptibility

Susceptibility to alcohol's effects on growth and morphology depends, in part, on genotype. An early mouse model of FAS demonstrated that a strain with lower ADH activity was more sensitive to ethanol than a strain with greater metabolic capacity (Chernoff, 1977). Decreased birth weights were observed among DBA mice but not among C57BL mice given similar doses (Yanai and Ginsberg, 1977). Human variability in genetic vulnerability has been observed in fraternal twins in which one was more severely affected (Christoffel and Salasky, 1975).

Paternal drinking

Since many women who drink heavily mate with men who drink heavily, the role of paternal drinking as a mediating factor must be considered. In the rodent, increased resorption of defective fetuses and behavioral anomalies have been reported in association with paternal exposure to ethanol, although low birthweights and morphologic abnormalities were not observed (Randall et al, 1982; Abel and Tan, 1988). To date, clinical studies of the association between paternal alcohol consumption and adverse pregnancy outcome have been limited. One study has reported an association between father's drinking prior to conception and decreased infant birth weight (Little and Sing, 1987).

PREVENTION

Increased understanding of the mechanisms which underlie alcohol's effects points to clinical intervention as a powerful strategy for the prevention of alcohol-related birth defects. In a study at Boston City Hospital, drinking patterns were identified at the time of registration for prenatal care (Rosett et al, 1983a). Women who reported drinking heavily were informed that they had a better chance of having a healthy baby if they abstained from alcohol use for the duration of their pregnancy. Two-thirds of the women who participated in counseling sessions reduced alcohol consumption before the third trimester. Growth retardation occurred significantly more often among infants born to the women who continued to drink heavily throughout pregnancy (Rosett et al, 1983b). Neonates born to women who drank heavily early in pregnancy and reduced consumption before the third trimester were comparable in weight, length, and head circumference to those born to rare and moderate drinkers. Abnormalities were identified most frequently among children born to women who continued drinking heavily. FAS was diagnosed in five children, all born to women who continued to drink heavily throughout pregnancy. These associations were independent of eight variables thought to influence fetal growth and development: maternal age, parity, ethnicity, cigarette smoking, marijuana use, prepregnancy weight, baby's sex and gestational age.

Reduction of heavy drinking has been associated with improved neonatal outcomes on growth parameters, somatic status, and/or CNS development in additional case reports and prospective studies. In a program in Stockholm, supportive counseling was provided to women who reported problem drinking at registration for prenatal care (Larsson, 1983). Among women who drank between 3 and 12 drinks a day, 100% reduced consumption. Of the women who drank more than 12 drinks a day, 78% reduced. Two babies were born to women who continued heavy drinking: both were growth retarded and one was diagnosed as FAS. Babies born to women who reduced did not demonstrate growth retardation.

In a program in Seattle, 80% of the women who drank heavily were able to modify their drinking patterns (Little et al, 1984). Benefits among their children were described as dramatic. Fetal alcohol effects occurred three times more often when mothers continued drinking than when mothers reduced drinking.

Counseling was associated with cessation of heavy drinking by 35% of the at-risk women in a program in Atlanta (Smith et al, 1987). Increased neurobehavioral alterations were found in infants of women who continued to drink during pregnancy as compared with abstainers and with women who discontinued use by mid-pregnancy (Coles et al,

1987). The observed benefits were attributed to cessation of drinking and not to maternal characteristics. No significant differences were found between continued and reduced drinkers in patterns of alcohol consumption reported at the time of clinic registration or in the prevalence of alcohol-related pathology, although the continued group had begun drinking at an earlier age.

Benefits have been observed to persist in older children as well as in neonates. Examinations conducted with 80 children whose mothers were abusive drinkers demonstrated that children of alcohol abusers had the highest incidence of growth retardation, mental retardation, behavioral disturbances and morphologic abnormalities, (Larsson et al, 1985). The median age of children was 22 months (range 18-27 months). The postnatal environment did not compensate for these alcohol-related birth defects. Children of women who reduced consumption did not differ from controls in physical development or behavior, but did display delays in speech.

Among another group of mother-infant pairs in Sweden, five children whose alcohol abusing mothers had stopped drinking in the twelfth gestational week exhibited normal growth, somatic status, and cognitive function at 10 and 18 months (Aronson and Olegard, 1985). Subnormal scores were reported among the nine children whose mothers continued heavy drinking throughout pregnancy.

Experimental studies confirm that an opportunity exists for physiologic restitution and modification of abnormalities which develop secondary to impaired growth (Anders and Persaud, 1980). Abnormal embryogenesis apparent on day 9 in rats exposed to high doses of alcohol on gestational days 5-8 was not apparent on day 20. Similarly the incidence of defects was much lower in rats exposed to acetaldehyde when examination occurred on the twelfth day rather than on the tenth, with no difference in resorption rates (O'Shea and Kaufman, 1981; Dreosti et al, 1981).

Alterations in neonatal brain size and structure may not be permanent. Cerebellar Purkinje cells revealed significantly smaller nuclei on day 7 among rat pups exposed to alcohol in utero (Volk et al, 1981). At days 12 and 17 there was no differences, suggesting that alcohol caused a developmental delay with a potential for recovery. Complete cortical lamination, apparent in controls at 14 days, was not observed in the alcohol exposed pups until 23 days (Jacobson et al, 1978).

The capacity for recovery is variable. In mice, cell necrosis affecting the neuroepithelium both in the closed caudal portion of the neural tube and in the unclosed cranial neural tube was observed six hours after ethanol exposure (Bannigan and Burke,, 1982). At 24 hours post-exposure, the neuroepithelium showed signs of repair. At 50 hours, the neuroepithelium was completely clear of debris resulting from cell necrosis, although some embryos had open neural tube defects and some had been resorbed. Many fetuses survived to day 19 with no apparent gross defects.

While these studies suggest that there is considerable capacity for compensatory growth and repair, further study is required to evaluate functional effects of developmental delays. Alcohol exposed brains may not be similar to controls on all neurologic, physiologic or behavioral measures.

The benefits observed following reduction in alcohol consumption are consistent with the hypothesis that FAS represents the cumulative result of adverse effects throughout gestation. Exposure at critical developmental stages will affect particular systems. Since alcohol has the capacity to adversely affect each stage of fetal development, the earlier in pregnancy that heavy drinking ceases, the greater is the potential for improved outcome. When heavy drinking ceases, abnormalities and growth retardation that develop in later stages will be prevented. In addition, maternal capacity to nurture improves when drinking stops.

The demonstrated benefits when heavy drinking ceases reinforces the value of providing supportive therapy to pregnant women at risk (Weiner and Larsson, 1987). Identification and treatment of problem drinking women holds great promise for the prevention of alcohol-related birth defects.

REFERENCES

Abel, E. L. and Sokol, R. J. (1987). Incidence of fetal alcohol syndrome and economic impact of FAS-related anomalies. *Drug Alc. Depend.*, 19, 51–70

Abel, E. L. and Tan, S. E. (1987). Effects of paternal alcohol consumption on pregnancy outcome in rats. *Neurotoxicol Teratology.*, 19, 51–70

Anders, K. and Persaud, T. V. N. (1980). Compensatory embryonic development in the rat following maternal treatment with ethanol. *Anat. Anaz.*, 148, 375–83

Aronson, M. and Olegard, R. (1985). Fetal alcohol effects in pediatrics and child psychology. In Rydberg, U., Alling, C., Engel, Pernow, B., Pellborn, L. A. and Rossner, S. (eds.), *Alcohol and the Developing Brain.* Raven Press: New York, 135–45

Bannigan, J. and Burke, P. (1982) Ethanol teratogenicity in mice: a light microscopic study. *Teratology.*, 26, 247–54

Chernoff, G. F. (1977). The fetal alcohol syndrome in mice: an animal model. *Teratology.*, 15, 223–30

Christoffel, K. K. and Salafsky, I. (1975). Fetal alcohol syndrome in dizygotic twins. *J. Pediatr.*, 87, 963–7

Coles, C. D., Smith, I. E., Fernhoff, P. M., and Falek, A. (1985). Neonatal neurobehavioral characteristics as correlates of maternal alcohol use during gestation. *Alcoholism Clin. Exp. Res.*, 454–60

Dreosti, I. E., Ballard, F. J., Belling, G. B., Record, I. R., Manuel, S. J. and Hertzel, B. S. (1981). The effect of ethanol and acetaldehyde on DNA synthesis in growing cells and on fetal development in the rat. *Alcoholism Clin. Exp. Res.*, 5, 357–62

Ernhart, C. B., Sokol, R. J., Martier, S., Morton, P., Nadler, D., Ager, J. W. and Wolf, A. (1987). Alcohol teratogenicity in the human: a detailed assessment of specificity, critical period, and threshold. *Am. J. Obstet. Gynecol.,* 156, 33–9

Fisher, S. E., Atkinson, M., Burnap, J. K., Jacobson, S., Sehgal, P. K., Scott, W. and Van Thiel, D. V. (1982). Ethanol-associated selective fetal malnutrition: a contributing factor in the fetal alcohol syndrome. *Alcoholism Clin. Exp. Res.,* 1982, 6, 197–201

Jacobson, S., Rich, J. and Tovsky, N. J. (1978). Delayed myelination and lamination in the cerebral cortex of the albino rat as a result of the fetal alcohol syndrome. In Galanter, M. (ed.) *Currents in Alcoholism.* Grune and Stratton: New York, 123–33

Larsson, G. (1983). Prevention of fetal alcohol effects: an antenatal program for early detection of pregnancies at risk. *Acta Obstet. Gynecol. Scand.,* 62, 171–8

Larsson, G., Bohlin, A. B. and Tunnell, R. (1985). Prospective study of children exposed to variable amounts of alcohol in utero. *Arch. Dis. Child.,* 60, 316–21

Little, R. E. and Sing, C. F. (1987). Father's drinking and infant birthweight: report of an association. *Teratology.,* 36, 59–65

Little, R. E., Young, A., Streissguth, A. P. and Uhl, C. N. (1984). Fetal alcohol effects: effectiveness of a demonstration project. *CIBA Foundation Symposium 105, Mechanisms of alcohol damage in utero.* Pitman Press: London, 254–74

Lochry, E. A., Randall, C. L., Goldsmith, A. A. and Sutker, P. B. (1982). Effects of acute alcohol exposure during selected days of gestation in C3H mice. *Neurobehav. Toxicol. Teratol.,* 4, 15–19

Majewski, F. (1981). Alcohol embryopathy: some facts and speculations about pathogenesis. *Neurobehav. Toxicol. Teratol.,* 3, 129–44

O'Shea, K. S. and Kaufman, M. H. (1981). Effect of acetaldehyde on the neuroepithelium of early mouse embryos. *J. Anat.,* 132, 107–18

Pierce, D. R. and West, J. R. (1986). Blood alcohol concentration: a critical factor for producing fetal alcohol effects. *Alcohol,* 3, 269–72

Pratt, O. E. (1980). Fetal alcohol syndrome: transport of nutrients and transfer of alcohol and acetaldehyde from mother to fetus. *Psychopharmacology of Alcohol.* Raven Press: New York, 229–56

Randall, C. L., Burling, T. A., Lochry, E. A. and Sutker, P. B. (1982). The effect of paternal alcohol consumption on fetal development in mice. *Drug Alcohol Depend.,* 9, 89–95

Randall, C. L. and Taylor, W. J. (1979). Prenatal ethanol exposure in mice: teratogenic effects. *Teratology.,* 19, 305–12

Rosett, H. L., Weiner, L. (1984). *Alcohol and the Fetus: A Clinical Perspective.* Oxford University Press: New York

Rosett, H. L. and Weiner, L., Edelin, K. C. (1983a). Treatment experience with pregnant problem drinkers. *J. Am. Med. Assoc.,* 249, 2029–33

Rosett, H. L., Weiner, L., Lee, A., Zuckerman, B., Dooling, E. and Oppenheimer, E. (1983b). Patterns of alcohol consumption and fetal development. *Obstet. Gynecol.,* 61, 539–46

Samson, H. H. and Grant, K. A. (1984). Ethanol induced microcephaly in the neonatal rat: relation to dose. *Alcoholism Clin. Exp. Res.,* 8, 201–3

Smith, I. E., Lancaster, J. S., Moss-Wells, S., Coles, C. D., and Falek, A. (1987). Identifying high-risk pregnant drinkers: biological and behavioral correlates of continuous heavy drinking during pregnancy. *J. Stud. Alcohol.,* 48, 304–9

Sokol, R. J. (1981). Alcohol and abnormal outcomes of pregnancy. *Can. Med. Assoc. J.,* 125, 143–8

Volk, B., Maletz, J., Tiedemann, M., Mall, G., Klein, C. and Berlet, H. H. (1981). Impaired maturation of Purkinje cells in the fetal alcohol syndrome of the rat: light and electron microscopic investigations. *Acta Neuropathol.,* 54, 19–29

Webster, W. S., Walsh, D. A., Lipson, A. H. and McEwen, S. E. (1980). Teratogenesis after acute alcohol exposure in inbred and outbred mice. *Neurobehav. Toxicol.,* 2, 227–34

Weiner, L., and Larsson, G. (1987). Clinical prevention of fetal alcohol effects—a realty. *Alcohol Health and Research World.,* 2, 60–63

Yanai, J. and Ginsburg, B. E. (1977). A developmental study of ethanol effect on behavior and physical dependence in mice. *Alcoholism Clin. Exp. Res.,* 1, 325–33

Part D

PREVENTION AND INTERVENTION

CHAPTER 7

Prevention Issues in Developing Programs

Darlind J. Davis, M.Ed.

INTRODUCTION

Prevention of alcohol and other drug problems has achieved national attention in recent years, unparalleled in media attention, community outcry, and government priority. Focusing primarily upon youth, America has become educated in the dynamics of addiction. Front page attention, network specials and public service messages abound with personal stories. The rich and famous, poor and destitute, together with families just like those next door attest to the tragedy of chemical dependency on the nightly news. But throughout this highly publicized "war on drugs" no target group has been more noticeably missing than women. Women are showing increases in heavy drinking, early onset of drug and alcohol use, and combined diagnoses which require special treatment, support and follow-up (Roman, 1988).

As women have risen to positions of influence in the community and workplace, the incidence of smoking, alcohol abuse and drug dependency has also risen proportionately. According to recent findings of the National Institute on Alcohol Abuse and Alcoholism, young persons under age 30 have doubled in self-help groups such as A.A. (Alcoholics Anonymous) since 1977. Women comprise one-third of the membership of A.A. Additionally, the following data illustrate the growing problem of alcoholism and drug addiction among women.

- 60% of all prescribed tranquilizers are intended for women.
- 71% of all antidepressant prescriptions are written for women
- 80% of all amphetamines are prescribed for women
- drug-related emergency room visits are likely to be a result of prescription drugs for women and street drugs for men.

Past studies have shown that men and women enter rehabilitation programs for different reasons. For example, women cite individual concerns for their drinking, such as relief from stress, to sleep better, to forget problems and pressures, and to settle nerves (Mulford, 1977). Further surveys of drinking practices show that in general "psychological dependence" is decreasing as a reason for seeking treatment, but women are showing much less of a decrease for this reason.

PREVENTION TARGET GROUPS

Dr. Elizabeth Morrissey, a noted researcher outlined the following profile of the woman who is most likely to drink, to be a heavy drinker, and use drugs/alcohol in order to reduce stress: she is 30 to 45 years old; employed, probably in a high status occupation; and is unmarried. Her drinking corresponds to the drinking patterns of men, and she tends to use alcohol to relieve stress and anxiety and to mute problems. Though she may seek treatment for her addiction problems at an earlier age than her male counterpart, the age of onset of these problems is much later. Upon entering treatment, her self-esteem is lower, and she is more likely to report feelings of powerlessness and inadequacy than a male (Shuckit, 1978). To compound her alcohol and other drug problems, she will also be more likely to have a history of suicide attempts, psychiatric counseling, and abuse of legally prescribed substances (Blume, 1986). She is likely to be more distressed by changes in sexual response resulting from chemical dependency than a male, and she may have a compounded problem of seeking rehabilitation with inadequate provisions for the care of her children. (Wilsnack, 1982).

In summary, this young woman is experiencing a complex interplay of social, medical and cultural forces. She has likely hidden her disease longer than average, she probably has small children and is without emotional support in the home due to divorce. She is emotionally bankrupt, at-risk for a variety of serious illnesses in addition to her alcoholism/drug dependency, and has fewer programs to choose from than a man of similar means (J. Johnson, 1988). Lack of programming in the areas of prevention, intervention and treatment continues to be problematic. While communities have made progress in recent years, the number of facilities specializing in women's addiction is limited.

Strategies to Prevent the Onset of Alcohol and Other Drug Problems

To plan effective strategies for women to prevent the onset of alcohol and other drug problems, the following implications must be considered:

- dual addiction must be addressed in materials produced for women. (45% of women report dual addiction at intake)

- biomedical findings on PMS, physical differences and effects of addiction on progression of other diseases have been underemphasized. New evidence shows differences in metabolism affect intoxication and menstruation (Sutker et al 1987).
- relationship of alcohol and other drug use to factors such as smoking, eating disorders, depression, AIDS, fetal alcohol effects, domestic violence, rape and incest.
- legally-obtained, as well as illegal substances, must be included in educational content.
- the affect of media and advertising on women, subtle messages in women's magazines which glamorize drinking without considering the negative consequences.
- the importance of self-empowerment and individual responsibility as part of a comprehensive approach.
- competency-based programs which challenge the participants and have the potential to reinforce a woman's personal attributes.
- culturally-specific models which promote social understanding and acceptance, (not stereotyping, fear and alienation).
- information on high risk characteristics such as genetic predisposition (inherited tolerance for alcohol and drugs) for women who are children of alcoholics.
- necessary adjunct services which increase participation and full completion of programs, such as child care, transportation, access and availability of services in neutral settings (ie. community colleges versus treatment centers).

Public health models have emphasized sustaining efforts to reduce consumption of alcohol, regulate advertising, increase the price and promote warning labels on the health effects of alcohol and tobacco products. While the primary concern is for the unborn, the health of women is also a very important goal of fetal alcohol and other drug education programs. Alcohol and other drug use during pregnancy can contribute to severe consequences for the mother and the infant.

Environmental factors play as critical a role in prevention planning as individual factors. Social acceptability has been a primary reason for the increase in consumption among women. As an example shrewd marketing techniques by the alcohol industry have targeted non-drinkers and light drinkers. These advertisements project women who consume alcohol as having power, success and sexual attraction. However, the ability to resist forms of cultural, family and group pressure are essential survival skills for women.

WOMEN WITH RISK FACTORS

Within the discussion of prevention for women occurs a natural breaking point between those programs aimed at the general population and for those at risk. For young adult women in the general population, programs are designed to build personal and interpersonal strengths, judgment, coping mechanisms, social resistance techniques and other proactive strategies. On the other hand, women who are at risk, require additional components which not only build life skills but also include thorough, accurate scientific information regarding risk factors. Abstinence is strongly indicated for groups such as: (a) children of alcoholics and substance abusers, (b) women with familial mental health problems such as depression, severe anxiety, phobias; (c) women whose husbands, siblings, close friends or relatives are heavy drinkers or users of drugs; (d) women who are planning to become pregnant or are already pregnant, and (e) recovering persons who have completed a treatment regimen and must maintain sobriety and remain drug-free. For these persons social use of alcohol and drugs is not an option. The challenge for the preventionist is to successfully convey this message without sermonizing, alienating, or placating women.

EARLY PREVENTION STRATEGIES

Earlier prevention curricula for women involved assertiveness programs through exposure to role play, self-expression experiences and interpersonal communication skills training. The importance of these training programs should not be underestimated. We have learned much from aftercare programs where emphasis is placed on keeping the newly recovering "drug-free." These programs successfully taught recovering women to resist pressures to resume alcohol or other drug use. Through the years, the success of these efforts has had important implications for both prevention program planners and outpatient counselors. Assertiveness training of the 70's became social resistance training of the 80's and now "resiliency development" for the 90's.

Examples of this type of preventive strategy:

1. *Design Your Own Life,* a state model prevention program (funded by the Pennsylvania Office for Drug and Alcohol Programs), located at St. Francis Hospital in Pittsburgh, its emphasis is on life transitions such as divorce, widowhood, and displacement. Prevention workers provide communication and listening skills, personal goal-setting, career development and group counseling. In partnership with a community mental health center, the program encourages women to express suppressed anger, talk out frustrations,

and provide meaningful support to others experiencing similar problems.
2. *Project WAIT (Wellesley Alcohol Informational Theater)* exemplifies a prevention program designed for college women, using live performances to act out critical issues. Wellesley College, long committed to building awareness on the effects of alcohol use upon women, has shared the model throughout the country.

CURRENT PREVENTION STRATEGIES

During the next phase of prevention programming, assertiveness training was combined with life skills education to create behavioral outcomes. Group change was emphasized as opposed to individual change. Environmental factors took on greater importance as it became evident that women cannot sustain long-term lifestyle change without sufficient external support mechanisms. Social changes that offered community support for women to avoid alcohol and other drug abuse include media campaigns, changes in advertising and regulation.

Examples of these techniques are: (a) the National Institute on Alcohol Abuse and Alcoholism (NIAAA) women and alcohol campaign targeting young women with multicultural backgrounds; (b) educating the pharmaceutical and alcohol beverage industries regarding negative stereotyping in lectures and films *(Killing Us Softly* and *Calling the Shots);* (c) restrictions in smoking areas of public buildings, warning labels, dram shop legislation and the abandonment of "ladies night" marketing techniques.

Programming has taken on a wider, more comprehensive direction in recent years. Building on greater public awareness and support for early intervention, preventionists have moved into true proactive efforts aimed at younger, non-using women. For example, the Girl's Clubs of America receive federal funding from the Office for Substance Abuse Prevention for a National Resource Center in Indianapolis, Indiana. This concept builds the capacity of an existing organization serving young girls at risk for drug abuse.

Woman-to-Woman is a privately financed program of the Association of Junior Leagues of America that informs women about the risks associated with alcohol and other drug use. Community coordination and involvement with college-age women through sorority affiliations are important organizing concepts in this model which contains beautifully designed and comprehensive materials.

Another project, *Camp Fire, Inc.* is developing a model drug and alcohol prevention curriculum for school-aged children in grades K–6.

Prevention can be found in a variety of settings: schools, workplace centers and service clubs. Each and every organization within the community must take responsibility for including prevention/education segments on an on-going basis. This "institutionalization" of prevention adds to the impact of a joint community message to discourage drugs, update adult knowledge about alcohol and drugs to increase their credibility and bring the issue into an open forum to counter denial, stereotyping and misinformation. Preventionists agree that the task must be a partnership, a joint venture in which total responsibility rests with no one individual, agency or family.

Women as Preventors: An Adult-Teen Partnership uses role-modeling and information in a prevention curricula. The program was developed by the National Board of the YMCA in cooperation with the National Clearinghouse for Alcohol and Drug Information (M. Mong, et al 1983). This manual for training of trainers uses adult mentors in a comprehensive, intense, skill-building experience to reduce vulnerability factors.

Women as Preventionists

Women play an important role in the prevention movement. They have less patience with intoxication, are more likely to intervene in situations where someone is out of control, and have, in general, more negative attitudes toward problem drinking. This information is summarized by Elizabeth Morrissey (1984) and further suggests:

a. women may appropriately be natural providers of prevention services (Wilsnack, 1982);
b. victims of alcohol and other drug abuse may be most likely to mobilize in support of control policies which reduce such problems (Moore and Gertein, 1981); and
c. women have been key lobbyists in drunk driving legislation, funding for prevention and adolescent treatment services, and efforts to confront the alcohol industry.

We need to revaluate our efforts to serve women and expand the resources for women's health education and social development. Women, especially our nation's young women, require our collective attention to this oversight.

ACKNOWLEDGEMENTS

A special Thanks to Mary L. Grady for editing this chapter, Angela Theophile for researching articles, Bonnie Horner and Brenda Gonder for transcribing the final document.

REFERENCES

Blume, S. (1986). Women and Alcohol (Journal of the American Medical Association, V. 256 (No. 11), 1469–1476.

Hallowell, M., The Association of Junior Leagues, New York, New York, 1988.

Johnson, E. M. (1987). Women's health: issues in mental health, alcohol and substance abuse. *Public Health Reports* 102 (4. Suppl): pp. 42–48.

Johnson, J. (1988). Advances in Women's Alcohol and Drug Programs Workshop, NASADAD Annual Meeting.

Kilbourne, J. (1988). Under the influence, Lordly and Dame, Cambridge.

Mong, M., Bower, S. and Johnson, F. T. (1983). Women as preventors, National Board of the Young Women's Christian Association (YWCA) New York, N.Y.

Morrissey, E. R. (1984). Of women, by women, or for women? Women and alcohol, health related issues, monograph 16, ADAMHNA, proceedings of a conference, pp. 226–259.

Mulford, H. A. (1977). Women and men problem drinkers: sex differences in patients served by Iowa's Community Alcoholism Services. Journal of Studies on Alcohol 38 (9): 1624–1639.

National Council on Alcoholism, New York, New York. Federal response to a hidden epidemic: alcohol and other drug problems among women.

Noble, J. A. (1988).

National Institute on Alcohol Abuse and Alcoholism, National data on women's alcoholism and alcohol abuse.

Roman, Paul M. (1988). *Women and Alcohol Use:* A Review of the Research Literature. Alcohol, Drug Abuse, Mental Health Administration, U.S. Department of Health & Human Services.

Schuckit, M. A., and Morrissey, E. R. (1976). Alcoholism in women; some clinical and social perspective with an emphasis on possible subtypes. Alcoholism Problems in Women and Children. New York: Grune and Stratton, 5–35.

Sutker, P. B., Goist, K. C. and King, A. R. (1987). Acute alcohol intoxication in women: relationship to dose and menstrual cycle phase. *Alcoholism: Clinical and Experimental Research* 11: 74–79.

Wilsnack, S. C. (1982). Alcohol abuse and alcoholism in women. In: Pattison, E. M., and Kaufman, Ed., eds. Encyclopedic Handbook of Alcoholism. New York: Gardner Press, pp. 718–735.

CHAPTER 8

Employee Assistance Programs

*Mary Ellen Lukina-Wiersma,
M.S.W., CEAP, C.A.C.*

As women continue to enter into the work force in increasing numbers, Employee Assistance Programs (EAPs) are in a unique position to offer assistance to women with alcohol and other drug problems. It has been estimated that as many as one half of the approximately 12 million Americans with drinking problems are women (Page, 1986, p. 50). Of the increasing number of cocaine abusers,, as many as half may be women. Twice as many women than men have a dual dependence on alcohol and anti-anxiety drugs (Agrest, 1987, p. 50). Given these statistics, chemical dependency among women in the workplace with their increased access to and use of chemicals should be a significant problem. Yet, experience and program data shows that the EAP counselor's case load does not reflect this.

There is a growing body of knowledge revealing ways in which chemical dependency is different for women. Understanding and applying this research could lead to more effective interventions in the workplace for chemically dependent women.

Women's drinking and drug use patterns have changed dramatically. Today over 60% of the women in the United States drink alcohol. Recent studies suggest that the amount of alcohol consumption of women is on the rise.

Among female college students, heavy drinking has more than doubled in the past eight years (Pape, 1986, p. 58). It is now known that women are more sensitive to alcohol, become addicted more easily, develop alcohol-related physiological problems and die sooner (Rovner, 1987, p. 1). Women's reaction to alcohol also varies day to day, in relation to her menstrual cycle.

Many alcoholic women also use illegal or other drugs. Cocaine use by women has increased. Barbara Cooper-Gordon, Administrator of New York's Stuyvedant Square's Chemical Dependency program says that "Women may not feel comfortable drinking in a bar for hours, but cocaine is a nice, neat drug—for women who want to deceive themselves" (Van Gelder and Brandt, 1986, p. 100). Dr. Josette Mondanaro explains that "if we look at what's being demanded of women in the 1980's, we begin to get an idea of why coke seems to be the perfect drug for women." Mondanaro goes on to say "the ideal female is thin and coke has a reputation as an appetite suppressant. She is supermom and coke has the reputation of providing boundless energy. Also, we live in a society that still tends to denigrate women and coke bolsters self esteem. Many report initial feelings of confidence, flirtatious, outgoing, thin, sensual and desirable" (Van Gelder and Brandt, 1986, p. 100). Cocaine is a powerful, addictive drug being used more by women to combat the demands facing them every day.

Along with the increased use of cocaine by women, widespread use of and substantial misuse and abuse of anti-anxiety agents and sedatives are largely ignored (Agrest, 1987, p. 52). Xanax, a drug many believed to be harmless is now understood to be addictive. The use of this drug has increased dramatically. Many women go to their doctor's with feelings of panic, anxiety, and depression. Valium or Xanax is prescribed adding to the effects of alcohol and potentially adding another addiction. Many women will readily deny the use of prescription drugs. Unfortunately, many women inappropriately use prescription drugs without the knowledge that they are addicting. Many doctors still view tranquilizers as the panacea for woman's problems. They continue to prescribe them not believing they are addicting and harmful. A woman may also switch doctors if the doctor refuses to renew her prescription.

This is information which seems to say that women are using chemicals at an alarming rate. One writer suggests that women who work tend to have higher rates of problem drinking and drug use then do homemakers (Pape, 1986, p. 50). This may be because the workplace creates pressures, unrealistic role expectations and role conflicts—placing the family and work priorities in direct competition (Clark and Covington, 1986, p. 8). Drinking and other drug use may be used as a coping mechanism for dealing with these stresses. The drug use could affect the woman's ability to perform her responsibilities at home and on the job.

Traditionally, EAPs have relied on supervisors to document a decline in job performance, an assumed reflection of chemical use and other problems, and refer employees to the program. Supervisors are trained in observing, documenting, and referring employees with job performance problems to the EAP. This presumes that supervisors should be intervening with the men and women who have chemical use problems. Women are not being referred as often as they should be to the EAP.

Significant decline in job performance in women may not be the only appropriate indicator of problems in women. Many women tend to occupy lower skill jobs and may be able to perform their job adequately under the influence of chemicals or hung over (Pape, 1986, p. 50).

Women in higher positions may be able to hide performance problems more easily, much like their male executive counterparts.

Also, some male supervisors have difficulty confronting women on job performance problems, fearing they will become emotionally distraught or attribute the job performance problem to "female problems" (Pape, 1986, p. 50). Therefore, even if a supervisor did indeed observe problem behavior, a discussion about the performance problem and subsequent referral most likely would not take place.

If indeed a performance discussion did take place, this alone may not be an adequate motivator to seek help. Data suggests that women are more likely to seek health care than men. In other words, women must feel more personally touched by the problem to seek help (Mastrich and Farmer, 1986, p. 10).

Discussing with the woman consequences beyond the actual performance of the job duties may be more effective. Noting how her poor performance could affect others, explaining what her role is, the importance of the role, and how her lack of contribution might hurt the whole group may be more effective. Personal concerns may be more of a triggering or motivating factor for women to seek help than confrontation (Mastrich and Farmer, 1986, p. 10).

It should be noted that other indicators may be more important and useful than focusing on the traditional job performance decline when intervening with a chemically dependent woman. Absenteeism is a much more significant indicator for women with chemical abuse problems. Due to the fact that women often are in more highly supervised positions, a woman would rather call in sick than risk the chance that her supervisor would see her intoxicated or experiencing a hangover. Other indicators of possible chemical dependency problems that could be included as job performance problems include: deterioration of personal grooming, frequent illness, impulsive vacation, and moody and irritable behavior (Pape, 1986, p. 51). Job performance problems need to be defined in a very broad sense, including ability to perform job duties and interact effectively with co-workers and superiors and subordinates along with appropriate use of sick and vacation time.

Even with a well trained supervisor discussing a broad range of performance problems, some chemically dependent woman may be more willing to quit her job than go through the embarrassment of being confronted by a supervisor and referred to the EAP. Obviously, formal supervisory referral is a lesser used avenue for women to reach the EAP. Women are much more likely to be self-referred or referred by a co-worker or family member. A broad brush EAP, versus an occupational alcoholism program, that strongly support self-referral and use by family members is possibly more effective in reaching the chemically dependent woman. Chemically dependent women tend to associate the onset of problem chemical use with a particular stressful event. Broad brush EAPs are available for all types of problems. If the EAP continues to promote the program actively stressing the broad nature of services offered, the women prone to drinking may avail themselves to the program when those stressful events take place.

When chemically dependent women do finally access the Employee Assistance Program, the process of assessment and referral is different than for men as women more readily ask for help with mental health or family issues and avoid any concerns around a chemical problem. EAP counselors must therefore understand this dynamic and assess appropriately. This means knowing that symptoms of a mental health or family problem can disguise the primary issue of chemical dependency. Denial of the chemical problem is very strong. Women who have chemical use problems are able to present very viable, serious problems, other than their chemical use that need immediate attention. This presentation of other serious problems may distract even the most astute employee assistance counselor away from the woman's own chemical use and focus on the other problems.

Women may not be trusting of an EAP program because of its connection to the workplace. Confidentiality of the program needs to be stressed. When a woman does indeed come to the EAP, the information about the drinking or drug use may come very slowly. Interpersonal connections based on friendship and trust are often more powerful than a coercive influence in attracting the woman to seeking treatment (Mastrich and Farmer, 1986, p. 10). The EAP counselor may need to move slowly, building trust with his/her assessment of a chemically dependent woman. Women appear to need much support in accepting that they are chemically dependent and seeking treatment for the disease.

There are many barriers unique to women when contemplating treatment that the EAP counselor needs to be aware of. If the woman is seeking inpatient treatment, the fear of losing her children and the guilt of not being available to her children may be overwhelming. For many women, this could be too strong a force to deal with and they then refuse to seek treatment. Even outpatient treatment may not be an option because of the lack of child care, especially for a single parent. Also, many women may not have the vacation or sick time to take off for treatment. Her employer may terminate her and find someone to replace her rather than find other options to holding the job open or be flexible with her schedule while she seeks treatment. EAP counselors must strive to show women the negative consequences her continued alcohol or drug use will have on her family. By trying to show treatment as a way of improving the quality of her family life, the counselor may be able to motivate the women into seeking help. Striving to help the women find alternative child care and seek treatment is an important process when working with women who are chemically dependent. Also, the task is to find treatment resources that deal specifically with women's issues. Treatment programs dealing with women's issues will aid in successful treatment and recovery. Child care may be one of the most important barriers to combat when dealing with chemically dependent women.

Insurance coverage becomes a concern because in many cases, their insurance may be lacking or inadequate in coverage for any type of chemical dependency treatment. EAP counselors must find funding sources and treatment centers working with women who have funding sources to help women who financially cannot afford treatment.

Experience has shown that in most cases, women who are chemically dependent have grown up in an alcoholic or dysfunctional family and/or have a spouse or partner who are actively chemically dependent. This adds an additional barrier to seeking help. There is no encouragement from the women's support group at home to seek treatment. In fact, many times the woman is introduced to the drug by a man. In the case of cocaine, "the man introduces the cocaine and it becomes the fulcrum of an oppressive relationship. The relationship gets cemented by the cocaine and the woman becomes dependent. The man is keeping them supplied or not supplied, he's in control and they'll accommodate themselves to whatever he wants to get the drug." (Van Gelder and Brandt, 1986, p. 101). A woman needs additional support to seek help to break away from this oppressive situation. It may take several visits with the woman before a successful referral takes place.

Alcohol abuse is closely associated with domestic violence, both on the part of the abuser and the victim, with the latter particularly unwilling to seek treatment in fear of the reprisals (Rovner, 1987, p. 1). Safety for herself and her children becomes an important issue that needs to be dealt with in the referral process. A woman may not be able to contemplate treatment until she is certain that her children are safe.

Once a woman completes treatment, reentry into the workplace can be very beneficial to the continued sobriety and recovery of a chemically dependent woman. Even though the workplace may originally have been a contributing factor to the frustration and role conflict that lead to the chemical abuse. Being at home, with little structure may be too stressful for a woman still learning to be comfortable with herself and her sobriety. One's career or employment is a major source of one's self concept in our society. Women need to view employment as positive and search for a way to feel good about the work that they are doing, therefore improving their self concept and self esteem. Work may be a way of alleviating anxiety, particularly because it structures time. Work can be a valuable source of support and structure for a recovering chemically dependent woman.

Employee Assistance Programs must continue to expand and redefine its role when working with chemically dependent women. Time and effort to move slowly with the assessment is necessary for a successful referral. There still appears to be considerable denial that women become chemically dependent. Introducing information about chemical dependency into many different forums, such as printed materials and trainings about the stresses of dual roles or seminars on parenting, will continue to raise the consciousness that women become chemically dependent and need help. To continue to attempt to intervene with women in a confrontation model based on job performance decline will not reap the results that EAP programs desire.

EAPs are in a unique position to impact many women's lives. Trust and support, sensitivity to women's issues, and knowledge of chemical dependency as it manifests itself in women may be more useful in reaching the chemically dependent women. Employee Assistance Programs are in the position to continue to educate and help working women and their families with chemical health issues.

REFERENCES

Agrest, Susan, (1987, October). Just a harmless little habit. *Savvy*. pp. 51–56.

Clark, A., and Covington, S. (1986, January). Women, Drinking and the Workplace. *The Almacan*. pp. 8–9.

Mastrich, J., and Farmer, J. L., (1986, September). The "Systems" Approach to Intervention. *The Almacan*. pp. 10–11.

Pape, Patricia, (1986, October). Women and Alcohol: The Disgraceful Discrepancy. *EAP Digest*. pp. 49–53.

Rovner, Sandy, (1986, May). Alcohol and Women: The Hidden Addiction. *Washington Post Health*. pp. 1–2.

Van Gelder, L., and Brandt, P., (1986, November). Women and Cocaine. *McCall's*. pp. 99–102.

CHAPTER 9

Adolescent Women
Gail Gleason Milgram, Ed.D.

INTRODUCTION

Alcohol is a part of American society. It is a socially acceptable legal commodity which is consumed by the majority of Americans. Approximately 2/3rds of the adult population drinks at some time during the year (U.S. National Institute on Alcohol Abuse and Alcoholism, 1987). Gallup's recent survey shows that 72% of the males indicate that they are consumers and 62% of the females (1985). Drinking is integrated into many activities and events of American life. It is possible to consume alcohol during a business lunch, a social gathering, dinner in a restaurant; it is also possible to drink while boating, bowling, or attending a sports event. Alcohol consumption is part of many family activities as well; religious ceremonies, parties and meals often serve as occasions of use. Though the children at home often participate directly or indirectly in these events, alcohol's role is usually not discussed. However the young are aware of what is consumed, to what extent, when and for what reasons. The lack of discussion reflects society's ambivalent attitudes and lack of clarity toward the use of alcohol. It also may stem from the misinformation and general discomfort which often surround the topic of alcohol. Since the majority of adults have had little alcohol education and are themselves drinkers, they are often unsure of the reasons for discussing the topic and the content to be presented.

Our adolescents grow and develop in a society which uses alcohol to be social, relax, enjoy special occasions, etc., while at the same time it is relatively unclear regarding our feelings and attitudes toward its use. We often get caught up in the debate of whether alcohol is good or evil, rather than in a discussion of what is appropriate and responsible and what is inappropriate and problematic related to alcohol use. Alcohol use all too often becomes an emotional and controversial topic. Since it is frequently masked in a lack of awareness and misinformation, it is often difficult to discuss the introduction of alcohol to the young.

Adolescent drinking is not limited to what they do when they're with other adolescents, it also includes activities that take place at home with parents and guardians (Milgram, 1982). Most young people are introduced to alcohol between the ages of 10 and 15 with the mean age being 13 (Blane and Hewitt, 1977). This introduction usually takes place with parents at home for the reasons the parents are drinking (beverage with a meal, family celebrations, religious ceremonies). Parents are often not aware that the introduction to drinking has occurred and, even if awareness is present, often don't discuss the issue with their children. This creates an information void for our young and also often makes it difficult for them to talk with each other about why they drink and how they feel about drinking.

YOUTHFUL DRINKING

Though drinking often begins at home in family situations, it also becomes integrated into adolescent activities which occur in other locations without parental supervision (e.g., teenage parties). This is an understandable occurrence as adolescents are searching for independence and establishing their own norms of behavior. The percentage of male and female adolescents who have indicated that they are drinkers is identified in select studies of high school students conducted in a variety of locations during the past 35 years.

The findings indicate that the majority of adolescents consume beverage alcohol, a phenomena that has remained constant during the time period represented by the studies. It is also apparent that more males identify themselves as drinkers, though in many studies the reported difference is quite small (86 to 85, 88 to 84, 82 to 80) and in others, it is somewhat larger (81 to 66, 88 to 75). However, males tend to drink more and more frequently than females; daily use of alcohol is reported by 6.4% of the young men and 3.1% of the adolescent females (Johnston, 1987). Though reasons for adolescent drinking are similar to adult reasons (e.g., to relax, be social, have a good time), drinking to intoxication is a more accepted outcome of a drinking experience for adolescents than for adults. Findings from the National Adolescent Student Health Survey (N. A. S. H. S.) indicate that approximately one-third of the 10th grade students report having had five or more drinks at least once in the two weeks prior to the survey. Fifty-eight percent of the males and 34% of the females in the survey of high school seniors, conducted by Johnston, et al, report consuming 5 or more drinks during a drinking occasion (1987). From the quantity of alcohol consumed, it can be assumed that many of these young people were intoxicated and in high-risk situations. Nineteen percent of the teen-age males and 9%

	% of Drinkers	
	Males	Females
Nassau County NY (Hofstra Research Bureau, 1953).	86	85
New York State (Mandell et al, 1963).	81	66
Oregon (Demone, 1966).	88	84
South Carolina (Kimes et al, 1969).	88	75
San Mateo County (Blackford, 1975).	82	80
New York City (Lee et al, 1975).	80	75
U. S. A. Nationwide (Rachel et al, 1975).	77	69
U. S. A. Nationwide (Johnston et al, 1979).	94.4	91.9
New York State (Barnes, 1984).	74	68
U. S. A. Nationwide (Johnston et al, 1987).	92.3	90.6

of the females polled by the Gallup Organization indicated that they had driven a car after drinking; 35% of the adolescent males and 36% of the females responded that they had been a passenger in a car when the driver was under the influence (1985). However, it must be noted that the response for females between the ages of 16–18 years regarding being a passenger with a driver who was under the influence was 54%.

These statistics identify three important concerns regarding adolescent female drinking. The first is the population who are drinking daily (3.1%) and may be dependent on alcohol; the second is the one-third (34%) of the adolescent females who on occasion drink to intoxication and therefore place themselves in high risk situations; and the third is the 9% of the females who indicate driving after drinking and 54% of the 16–18 year old females who indicate that they have been a passenger in a car with an intoxicated driver. Social activities also pose problems for adolescent females. A great deal of the socializing takes place at parties without adult supervision; alcohol, and other drugs, are often part of the scene. Gomberg notes that added concern for females in this situation is based on anxiety about sexuality and reproduction (1989).

Many of the patterns and concerns regarding alcohol consumption are also apparent through the college years. In their study of college drinking in the early 1950's, Straus and Bacon found 80% of the male and 61% of the female college students were drinkers (1953). In a study conducted almost thirty years later, Wechsler and McFadden indicated that 95% of the students were drinkers (1979). Engs and Hanson's recent survey of colleges throughout the U. S. A. found that 78.8% of the students drank (1988). College drinkers can be categorized as light, moderate and heavy; and drinking occasions can be discussed as infrequent, regular and daily. For the most part, heavy drinking is limited to week-end parties; however, it must be noted that Thursday begins the week-end on many campuses. Students often try not to schedule Friday classes so that they can fully participate in the parties. Reasons for drinking in college are similar to those identified by the high school population (e.g., to feel good, to be part of the group) with an even greater emphasis by some on drinking to get drunk. Being drunk seems to be considered a normal consequence of the drinking experience by many college students whether they engage in the practice or not. A recent survey of college students found that 56% of the respondents felt that college students drank too much (Gallup, 1985). Thirty-five percent of the college males surveyed by Gallup and 21% of the females responded affirmatively to the question "do you sometimes drink more than you should?" (1985).

Statistics on drinking among female college students identify areas of concern. The majority of females indicate that they are drinkers and many respond that they drink to enhance social interaction. However, 21% feel that they drink more than they should and approximately 25% experience intoxication four or more times a month. Intoxication is related to passing out for about 10% and for hang-overs for about one-third of the females. Another serious consequence of intoxication for about 20% of the females is that they don't remember what happened when they were drinking. Intoxication negatively affects a percentage of female college students and puts them at-risk to experience problems.

ALCOHOL EDUCATION

Youthful drinking is a reality which elicits societal concern. Since our schools are considered a viable source of information and help for societal problems, educational programs have evolved to address this issue. Unfortunately, many have lacked a comprehensive approach and integration into the school's curriculum. They are often one-shot, one-program events that receive publicity to calm the public; however, they do little to help young people understand alcohol use and its impact on their lives. Alcohol education should span the public school years (K–12) and also be an integral part of programming on the college campus. The institution's policy should state when alcohol will be covered in the curriculum, what department will have responsibility for this task, and indicate who will be providing the education (Milgram, 1987). The philosophy on which the program is based should also be a part of the policy. This enables a clear statement of program goals and expected outcomes (Milgram and Griffin, 1985). In this way, the community will understand the various aspects of the program and how they fit together. Parents will also be able to discuss the philosophical position with their children. There is a major difference in the responsible decision-making strategy and the "just say no" approach. Since the content and techniques emanate from the program's philosophy, it is essential that the support structure be understood by all concerned. All of our children and adolescents, both males and females, should receive information regarding alcoholic beverages, why people drink, the effects of alcohol on the human body, problems related to the use of alcohol, the disease of alcoholism, and the impact of alcoholism on the family and society; specific information dealing with female drinking should also be addressed (patterns of use, relationship to menstrual cycle, Fetal Alcohol Syndrome, etc.). Information on drinking and driving and on being a passenger in a car with an intoxicated driver needs to be discussed with both genders. Due to the high rate of females who indicate that they have been a passenger in a car with an intoxicated driver, this topic must be stressed with female students. The relationship of alcohol to sexuality and the phenomena of date-rape also need to be highlighted.

Techniques and strategies for alcohol education include self-esteem and image building, refusal skills, assertiveness training, decision-making, role plays, and discussion exercises. To incorporate the variety of strategies into the program and to handle the range of content suggested, teachers need to be trained. They also require training on handling specific issues related to female drinking. Young male drinking seems to be accepted as traditional, masculine behavior but young female drinking is a more sensitive issue (Gomberg, 1989). This may result in parents being more uncomfortable in discussing drinking and intoxication with females, especially because of the implications of alcohol use to sexuality.

The educational institutions, kindergarten through college, need to develop and support students who require assistance. The policy statement should indicate the availability of a Student Assistance Program (SAP). Though all student problems are not related to alcohol, approximately 50% of them are related to the students' alcohol/drug use or that of a family member (Milgram, 1988). Students should know how the program can help, understand that it is confidential, and know the ranges of support services that the SAP can provide. It is also important that educational institutions develop employee assistance programs (EAP's), as educators experience alcohol/drug problems, have spouses who are alcoholics, or are themselves children of alcoholics. Other problems requiring assistance also exist in this population and help and support should be available. Though EAP's are growing in corporate America, school systems and colleges have been slower to recognize the needs of their employees.

IMPLICATIONS

The information on youthful drinking and alcohol education has implications for information programs, educational efforts, intervention techniques, and treatment for females. The following items are offered as a starting place to impact on policy issues and program development for females:

- A majority of adolescent and adult females indicate that they are drinkers; therefore, alcohol information and educational programs directed toward women need to be developed.
- Females experience greater intoxicating effects of alcohol due to differences in body water and fatty tissue than males, a fact which should be understood by drinking females.
- Females need to be aware of the differences as well as the similarities of alcohol use of males and females.
- Motivations and reasons for drinking by females should be discussed.
- Supportive strategies need to be identified to help females reduce risks related to alcohol use (e.g., not being a passenger in a car with an intoxicated driver).
- The critical importance of transition periods (high school to college) and their relationship to alcohol need to be acknowledged (Berkowitz and Perkins, 1987).
- The double standard regarding female drinking, especially when it results in intoxication, needs to be discussed.
- Awareness programs on alcohol's relationship to sex, AIDS, and date-rape need to be promulgated.
- Intervention strategies to help females cope with alcohol issues and problems need to be developed.
- Counseling techniques and treatment methods which are sensitive to the needs of females should be strengthened. Specifically designed and structured treatment follow-up programs are also needed to enhance recovery for women (Roman, 1988).

REFERENCES

American Alliance for Health, Physical Education, Recreation and Dance (1989). *National Adolescent Student Health Survey.* U.S. Department of Health and Human Services: Washington, D.C.

Barnes, G. M. and Welte, J. W. (1984). *Alcohol Use Among Secondary School Students in New York State.* Research Institute on Alcoholism, Division of Alcoholism and Alcohol Abuse: Buffalo, NY.

Berkowitz, A. D. and Perkins, W. (1987). Recent Research on Gender Differences in Collegiate Alcohol Use: *Journal of American College Health.* 36(2) 123-129.

Blackford, St. Clair, L. (1975). *Student Drug Use Surveys—San Mateo County California, 1968-1975.* San Mateo County Department of Public Health and Welfare: San Mateo, CA.

Blane, H. T. and Hewitt, L. E. (1977). *Alcohol and Youth; An Analysis of the Literature, 1960-1975.* National Technical Information Service: Springfield, VA.

Demone, H. W. (1966). *Drinking Attitudes and Practices of Male Adolescents.* Ph.D. Dissertation, Brandeis University: Boston, MA.

Engs, R. C. and Hanson, D. J. (1988). University Students' Drinking Patterns and Problems: Examining the Effects of Raising the Purchase Age: *Public Health Reports.* 103(6) 667-673.

Gallup, G. (1985). *Alcohol Use and Abuse in America,* Gallup Report #242. Gallup Poll: Princeton, NJ.

Gomberg, E. L. (1989). *Alcohol and Women.* Center of Alcohol Studies, Rutgers University: New Brunswick, NJ.

Gusfield, J. R. (1961). The Structural Context of College Drinking: *Quarterly Journal of Studies on Alcohol.* 22 428-443.

Harford, T. C., Wechsler, H. and Rohman, M. (1983). The Structural Context of College Drinking: *Journal of Studies on Alcohol.* 44(4) 722-731.

Hofstra Research Bureau, Psychological Division, Hofstra College (1953). *Use of Alcoholic Beverages Among High School Students.* Sheppard Foundation: New York, NY.

Humphrey, J. A. and Friedman, J. (1986). The Onset of Drinking and Intoxication Among University Students: *Journal of Studies on Alcohol.* 47(6) 455-458.

Johnston, L. D., Bachman, J. G. and O'Malley, P. M. (1979). *Drugs and the Nation's High School Students.* National Institute of Drug Abuse: Rockville, MD

Johnston, L. D., O'Malley, P. M. and Bachman, J. G. (1987). *National Trends in Drug Use and Related Factors Among American High School Students Young Adults, 1975-1986.* National Institute on Drug Abuse: Rockville, MD

Kimes, W. T., Smith, S. C. and Maher, R. E. (1969). *Alcohol and Drug Abuse in South Carolina High Schools.* Department of Education: Columbia, SC.

Lee, E. E., Fishman, R. and Shimmel, G. M. (1975). *Emerging Trends of Alcohol Use and Abuse Among Urban Teenagers.* Paper presented at Annual Conference of the National Conference of the National Council on Alcoholism: Milwaukee, WI.

Mandell, W., Cooper, A., Silberstein, R. M., Novick, J. and Koloski, E. (1963). *Youthful Drinking, New York State 1962.* Report to the Joint Legislature Committee on the Alcoholic Beverage Control Law of the New York State Legislature: Staten Island, NY.

Milgram, G. G. (1987). Alcohol and Drug Education Programs: *Journal of Drug Education.* 17(1) 43-56.

Milgram, G. G. (Oct. 28, 1988). *Impact of a Student Assistance Program.* Paper presented at Smithers Prevention Symposium: New Brunswick, NJ.

Milgram, G. G. (1982). Youthful Drinking: Past and Present: *Journal of Drug Education.* 12(4) 289-308.

Milgram, G. G. and Griffin, T. (1985). *What, When and How to Talk to Students About Alcohol and Other Drugs: A Guide for Teachers.* Hazelden: Center City, MN.

Mills, K. C. and McCarty, D. (1983). A Data Based Alcohol Abuse Prevention Program in a University Setting: *Journal of Alcohol and Drug Education.* 28(2) 15–27.

Rachel, J. V., Williams. J. R., Brehm, M. L., Cavanaugh, B., Moore, R. P. and Eckerman, W. C. (1975). *A National Study of Adolescent Drinking Behavior, Attitudes and Correlates.* Research Triangle Institute, Center for Study of Social Behavior: Research Triangle Park, NC.

Ratliff, K. G. and Burkhart, B. R. (1984). Sex Differences in Motivations for and Effects of Drinking Among College Students: *Journal of Studies on Alcohol.* 45(1) 26–32.

Roman, P. M. (1988). *Women and Alcohol Use: A Review of the Research Literature.* U.S. Department of Health and Human Services: Rockville, MD.

Saltz, R. and Elandt, D. (1986). College Student Drinking Studies 1976–1985: *Contemporary Drug Problems.* 117–159.

Straus, R. and Bacon, S. D. (1953). *Drinking in College.* Yale University Press: New Haven, CT.

U.S. National Institute on Alcohol Abuse and Alcoholism (1987). *Sixth Special Report to the U.S. Congress on Alcohol and Health.* U.S. Government Printing Office: Washington, D.C.

Wechsler, H.. and McFadden, M. (1979). Drinking Among College Students in New England: Extent, Social Correlates and Consequences of Alcohol Use: *Journal of Studies on Alcohol.* (40) 969–996.

CHAPTER 10

College Women[1]
Ruth C. Engs, Ed.D., R.N.

DRINKING PATTERNS AND PROBLEMS

Alcohol is the major drug problem on the college campus. Studies over the past 30 years have shown that female college students are less likely to drink and to get into problems with their drinking compared to male college students. However, they have been drinking more frequently and in greater quantities since the late 1970s compared to the 1950s and 1960s. During this present decade many women students have been found to exhibit abusive drinking patterns and alcohol related problems. Drinking has also been shown to be a contributing factor in rape and pregnancy. Women students are more likely than men to be put into circumstances where their date may be too intoxicated to safely drive them home (Engs and Hanson, 1985; Engs and Hanson, in press Berkowitz and Perkins, 1987).

Some recent data collected by David Hanson, Sociology, Potsdam, NY and myself show a decrease ($X^2 = 22.0$ df $= 2$ p $< .001$) in the percent of female students drinking once a year or more since 1985. A sample of about 2000 female students in each of three time periods from the same 56 colleges in all areas of the United States was compared. The results show that 80% of the students drank during both the 1982–83 and the 1984–85 academic year compared to 73% in the current 1987–88 academic year. This same trend has been found in males and may be the result of raising the drinking purchase laws. This decrease resulted from fewer light and moderate drinkers. (Engs and Hanson, 1988).

Though there has been a decrease in the percent who drink, there has been no change in heavy drinking among women over this decade. Nor has there been a change in 13 of 17 alcohol abuse behaviors. Other than for drinking and driving problems which have continued to decrease

---

1. Based upon paper presented at Alcohol is a Women's Issue Conference. Junior League/National Council on Alcoholism, September 29–30, 1988, Washington, DC.

TABLE 10.1. Percent Female College Students Exhibiting Problems Related to Drinking in the Past Twelve Months. Sample Size: 1982–83 = 1974; 1984–85 = 1685; 1987–88 = 2021.

	1983	1985	1988
Hangover	70	70	73
Vomiting	**40**	**43**	**47***
Drive Car After Drinking	**50**	**47**	**43***
Drive Car After Knowing Had Too Much to Drink	31	30	30
Drink While Driving	**36**	**32**	**29***
Come to Class After Drinking	6	6	7
Cut Class Because of Drinking	6	8	7
Missed Class Because of Hangover	22	25	24
Stopped for DWI	1	1	1
Criticized by Friend for Drinking Too Much	**8**	**9**	**12***
Trouble with the Law	2	2	3
Lower Grade	4	4	4
Problems with School Administration	1	1	2
Gotten Into Fight Due to Drinking	8	8	10
Think Have Problem with Drinking	7	6	7
Damage To School Property Due to Drinking	4	4	5
Lost Job Because Of Drinking	1	1	1
Heavy Drinking (5 drinks at one sitting once a week or more)	15	15	16

* $p < .01$

since 1982, there has been an increase in vomiting and being criticized for drinking too much (Table 1).

When the total sample of women students in the 1987 national sample (n=5280) were compared by race, it was found that a much higher percent ($p < .001$) of white students drank and were heavy drinkers compared to black students. There were not enough Asian, Hispanic or Native American students for comparison in this sample. Over 80% of white students drank at least once a year compared to only 51% of black women students.

Among students who drink once a year or more, a significantly (p .001) higher percent of white females (14%) are heavy drinkers (5 or more drinks at any one sitting) compared to black females (2%). For 12 of the 17 problems related to drinking, a significantly higher (p .001) percent of white students exhibited problems. There was no difference between the groups for missing class because of hangover, cutting class because of drinking, being criticized by a friend because of drinking, getting a low grade, problems with school administration and causing damage to school property because of drinking.

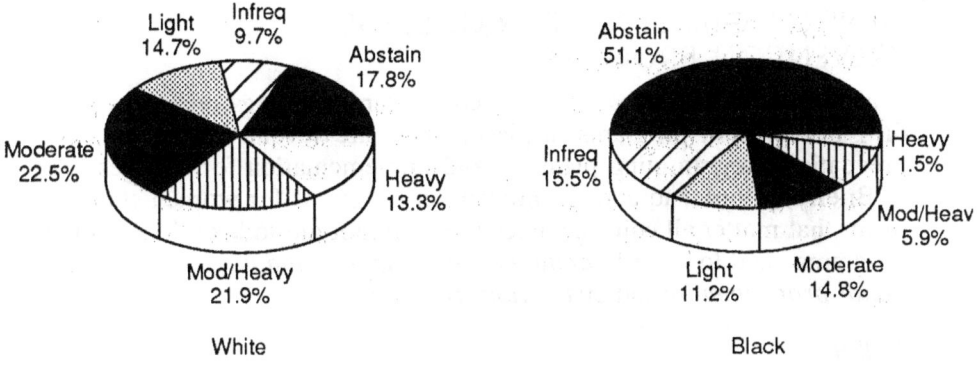

Light = drinks > once month and < 3 drinks   Infrequent = drinks > once year < once month
Moderate = drinks at least once week < 4 drinks each sitting
Mod/H = drinks 3-4 drinks once week   Heavy = > 5 drinks at least once week at one sitting

Figure 10.1. Comparison of White and Black Women College Students in Terms of Their Drinking Patterns.

The implications of these and other studies are that women college students, and in particular white students, do exhibit alcohol abuse problems which could contribute to serious problems that should be addressed by colleges.

OTHER DRUGS

Few studies have been accomplished concerning other drug use among college students. Johnson, O'Malley and Bachman's (1986) national study indicate that the annual prevalence rates for females for the following substances is as follows:

Annual Prevalence for Various Illicit Drugs for Women College Students

Marijuana	37%	Stimulants	11%
LSD	2%	Heroin	1%
Cocaine	15%	Barbiturates	1%
smoking (monthly rate)	26%	Opiates (other than Heroin)	1%

It appears that after alcohol, marijuana and cocaine are the most frequently used drugs by college women.

GENERAL GUIDELINES FOR COLLEGE PREVENTION PROGRAMS

Colleges and universities which receive federal fundings must now have drug and alcohol programs in place. There are several aspects of good prevention programming planning and implementation for colleges.

Briefly, in order to address prevention programs for women students, or for that matter all college students, it is important to keep in mind that programs need to have 1) *administrative support and policies,* 2) *educational programming* and *evaluation aspects.*[2]

Policies

Policies include administrative support, community support, college regulations concerning drinking on campus, and student assistant programs which can include peer counseling and education efforts.

Administrative support functions include a campus coordinator, planning committee composed of students, faculty, staff and community representation, Budget/Funding/Grant mechanisms and Student clubs and self-help groups. It would also include: residential life programming; counseling and medical facilities, campus security training and student affairs intervention and corrective action.

Community support organizations would include the local medical facilities, mental health facilities, law enforcement, religious organizations, taverns/restaurants, alcohol retailers, and self help groups such as AA, SOS, ACOA, AL-ANON, etc.

College regulations and policies concerning alcohol would include drinking policies at school sponsored parties and events, athletic events, academic functions, residential units/Greek, and dining facilities. It would also encompass disciplinary actions for intoxicated students, intervention procedures and treatment.

Educational Programming

Educational programming includes education for self responsibility, the physiology of alcohol, co-dependency, signs and symptoms of problem drinking and where to get help.

Education for self responsibility includes such items as wine appreciation, mixing drinks, steps for mature and safe alcohol use, alternatives to drinking, assertion and self-esteem building skills, stress reduction skills and tips for being a good host/hostess.

---

2. Based upon paper presented at St. Mary's College, Notre Dame, IN, November, 1979, "Alcohol Education for College Students."

The physiology of alcohol is also important and should include acute reactions to alcohol and the chronic effects of alcohol including alcoholism and problem drinking.

Problem drinking and alcoholism with its signs and symptoms, risk factors, and treatment.

Co-Dependency (adult children of problem drinkers) as a risk for problem drinking would include signs and symptoms, therapy and self help groups and referral sources.

DATA COLLECTION FOR NEEDS ASSESSMENT AND PROGRAM EVALUATION

To determine if prevention programs and campus policies have been effective in decreasing alcohol abuse and increasing knowledge concerning drinking, evaluations must be accomplished. However, baseline information first needs to be collected before programs are instituted. Both direct observation and indirect surveys should be accomplished.

Direct observation is the collection of data from various campus and community services. These would include the number of individuals over a period of time seen for alcohol, and or other drug related problems by the campus and local police, the community hospital and health services, legal services, student personnel office, and counseling centers. The number of arrests for DWI, incidence of vandalism, loud parties, fights and other violence related to alcohol abuse needs to be tabulated as part of this data base. However, since these types of data are often difficult to collect they are less likely to be gathered compared to indirect information.

Indirect information includes surveys of students, faculty or staff as to their drinking patterns and knowledge concerning alcohol, or other drugs, or perceived perceptions of drinking on campus. To determine students' self reported drinking, it is suggested that a simple reliable instrument, such as the *Student Alcohol Questionnaire,* be administered to a random segment of students on campus. This questionnaire has been used by hundreds of colleges over the past 10 years to determine drinking knowledge and behavior of college students. It was developed by the author and can be obtained free by contacting her.

After the problems have been assessed for a particular campus, policies, prevention and education programs, student assistance and other intervention programs can be instituted to address these needs.

Identification of High Risk Students for Prevention Efforts

For conserving limited prevention and education resources on college campuses, it is important to identify individuals who are at risk for alcohol abuse problems. It is thought that children from problem drinking homes have a higher probability of getting into drinking problems or, in particular if they are female, dating or marrying males who are likely to become problem drinkers. Freshmen, students from Roman Catholic backgrounds and for whom religion is not important, are also, more likely to abuse alcohol.

To help these high risk females who are found to be co-dependent and or have some of other risk factors, specific programs can be developed by campus religious groups, freshman orientation committees or student personnel or counseling centers. These women can be encouraged to enter support groups or educational situations where possible emotional problems and the issues of co-dependency can be addressed. Co-dependency status of students can be identified by use of the 11 item CODE instrument (Engs and Anderson, 1988). The types of educational and intervention programs aimed at these high risk individuals will depend upon local campus situations.

Developing Education Material Using the Guidelines of the Health Belief Model

Telling women students that drinking is illegal if they are under 21 years of age and that drinking can cause alcoholism or fetal alcohol syndrome problems is not likely to be effective in preventing alcohol abuse (Goodstadt and Caleekal-John, 1984). In order to change health behavior, the health belief model suggests that educational material must be based upon actual problems which are likely to occur soon after consumption. The reasons for this is that the model specifies (Rosenstock, 1974) that in order to change a person's behavior the person needs (1) to feel personally susceptible to the health problem, (2) to believe the problem will cause serious personal harm, and (3) to believe there are actions the individual can take in which the benefits outweigh the costs.

Hangovers appear to have occurred to almost all female students, and vomiting has occurred with almost a half of them as found in our national study. If these alcohol abuse problems can be reduced, it can be assumed that more serious problems such as missing class because of hangover, or driving related problems would in turn be reduced. Because hangovers and vomiting problems are common, it is likely that students may feel personally susceptible to them, thereby, causing them to be more likely to take actions to minimize the immediate problem compared to other long range problems.

The following is an example of the **type of educational messages,** based upon the three primary components of the health belief model using the vomiting problem.

If you drink more than 6 drinks in a few hours you are likely to vomit. Vomiting is likely to turn off your date or boyfriend. You can prevent vomiting by drinking no more than one or two drinks per hour.

Keeping in mind that college women, particularly if they are from alcoholic families, are thought to be more likely to abuse alcohol and more likely to date males who are at risk for abuse alcohol, some specific educational strategies can be developed for these problems which reflect the guidelines of the health belief model. An example of a message reflecting this is as follows:

> No wonder my date seemed too great, he's just like my father. Even though I really like him he is now too drunk to drive me home and I'm afraid I could get hurt. Guess I'd better see if I can get someone else that can get me home.

Awareness of Your Campus

Before these types of educational messages, or for that matter, any type of prevention programs for women can be developed, the type of college in terms of public or private, religious or non-denominational or all female or co-educational needs to be determined. Depending upon the demographics of a college, varied programs may or may not be able to be developed. In general select from the many approaches, techniques, ideas and philosophies you read or hear about which appear to be applicable to your situation. Be aware of programs which say they have "the answer". Programs which work well for a small, female Catholic college may cause increased problems on a large, public, co-educational one. Furthermore, various combined efforts and continued focus upon the subject appear most likely to be effective than single approaches or methods.

SUMMARY

In summary, female college students do abuse alcohol. In any educational program aimed at women there must be broad general guidelines and support from the campus administration. High risk women for abuse or of dating alcohol abusing men need to be identified and be encouraged to get help if they are having problems. Specific messages aimed at these high risk women, and for all women, based upon common abusive behaviors can be developed based upon the health belief model. Prevention and education programs must be developed for individual campus situations.

REFERENCES

Berkowitz, A. D. and Perkins, H. W. (1987) Recent research on gender differences in collegiate alcohol use. *Journal of American College Health.* 36 123–129.

Engs, R. C. and Hanson, D. J. (in press). Gender differences in drinking patterns and problems among college students: A review of the literature. *Journal of Alcohol and Drug Education.*

Engs, R. C. and Hanson, D. J. (1988) University students' drinking patterns and problems: examining the effect of raising the purchasing age. *Public Health Reports,* 103 (6) 667–673.

Engs, R. C. and Hanson, D. J. (1985) The drinking patterns and problems of college students: 1983. *Journal of Alcohol and Drug Education.* 31(1), 65–83.

Engs, R. C. *Alcohol and other Drugs: Self Responsibility* (1987) Tichenor Pub. Group: Bloomington, IN.

Engs, R. C. and Anderson, D. S. *The CODE Manual.* (1988) Campus Alcohol Consultations: Washington, DC.

Goodstadt, M. S. and Caleekal-John, (1984) Alcohol education programs for university students: A review of their effectiveness. *International Journal of the Addictions.* 19(7) 721–741.

Johnson, L. D., O'Malley, P. M. and Bachman, Jerald G., (1986) *Drug Use Among American High School Students, College Students, and Other Young Adults National Trends through 1985,* National Institute on Drug Abuse, 5600 Fishers Lane, Rockville, Maryland 20857.

Rosenstock, I. M. (1974) Historical origins of the Health Belief Model. *Health Education Monographs,* 2(4) 328–335.

PART E

TREATMENT

CHAPTER 11

Issues in Alcoholism Treatment

Jacqueline Wallen, Ph.D., M.S.W.

INTRODUCTION

In 1987, approximately 24 percent of those who used alcoholism treatment services were women (NIDA, 1989). Because women are a minority in alcoholism treatment settings, concern exists that treatment programs may reflect the needs of men more than those of women (Roman, 1988). The idea that women have special treatment needs that are not being served by existing programs has not, however, received a great deal of support in the research literature.

One reason for the failure of existing research to support such a view may be that very few treatment outcome studies have focused on women. Harrison and Belille (1987) estimate that only 8 percent of the subjects in studies published between 1970 and 1984 are women. Even when studies have included women, findings for women have rarely been presented separately from findings for men (Annis and Liban, 1980).

A second reason why existing research contributes relatively little to our understanding of women's treatment needs has to do with a general shortcoming of alcoholism treatment outcome research. While research has shown alcoholism treatment to be effective overall, it has, for the most part, failed to demonstrate that particular alcoholism treatment strategies are more or less effective than others (NIAAA, 1987) either for men or women (Roman, 1988). This is because, in the past, researchers have tended to contrast treatment strategies to one another without taking into account possible differences among individuals in their treatment needs. Approaches that are effective with one individual may be less effective with another. Instead of asking which treatment methods are most effective, researchers are increasingly beginning to ask which treatment methods are most effective for what kinds of patients (NIAAA, 1987). Such research promises to clarify the unique treatment needs of women as well as other groups.

Although it has not yet provided definitive answers on whether women have special needs in treatment or on what these needs might be, research on alcoholism treatment has pointed to several issues of potential importance for women. This paper will discuss some of these issues and suggest directions for future research.

ISSUES

Access to Care/Barriers to Care

Little is known about the conditions that discourage or facilitate entry into treatment for women. This is primarily because virtually all of the research has been carried out on women who have already entered treatment and therefor have surmounted whatever obstacles they have encountered (Roman, 1988). To get a valid picture of the barriers to care that exist for women it is necessary also to study women who were never referred to treatment or who, though referred, never actually entered treatment.

Because women are less likely than men to be arrested for drunken driving, women do not reach treatment through the criminal justice system with anywhere near the same frequency as men (Roman, 1988). Approximately 90 percent of those entering drunken driver programs are men (Laign, 1987). Also, employee assistance programs may recognize problematic drinking patterns in men at an earlier stage than in women (Ibid). Family problems and pressures may motivate women to enter treatment, but family responsibilities may prevent women from seeking treatment (Roman, 1988, Beckman and Amaro, 1986). Lack of child care is one of the most frequently reported barriers to treatment for alcoholic women (Wilsnack, 1982).

Primary care may be especially appropriate setting for outreach to women (Laign, 1987; Blume, 1980). Combined data for males and females from the National Institute of Mental Health's Epidemiologic Catchment Area Study indicates that within any 6 month period, almost 70 percent of individuals with alcohol problems make at least one ambulatory care visit (Shapiro, Skinner, and Kessler, 1984). The figure for women alone is probably even higher than 70 percent, since surveys examining sex differentials have found that women typically make more health care visits than men (Verbrugge and Wingard, 1986).

Research on women who drop out of treatment suggests that women, more than men, may be motivated to enter or remain in treatment by particular services that a treatment facility offers (Beckman and Bardsley, 1986). A Treatment Planning Guide prepared by The National Institute on Alcohol Abuse and Alcoholism (NIAAA, 1986) suggests that successful outreach to women may depend on two factors: a comprehensive approach to women's needs at the time they are seeking treatment (e.g. child care, financial problems, marital problems, housing, transportation, legal problems, etc.) and an emphasis on issues of special concern to alcoholic women (e.g. special gynecological and obstetrical issues such as fetal alcohol syndrome, the potential dangers of combining prescription drugs and alcohol, etc.). Roman (1988) cites research showing the effectiveness of outreach efforts that work through community groups, that attempt to increase awareness of women's alcohol problems among health care professionals, and that offer comprehensive assistance with family adjustment and child care issues.

Heterogeneity of Women in Treatment

Women in treatment for alcoholism are not a homogeneous population. Harrison and Belille (1987), for example, have used Chemical Abuse-Addiction Treatment Outcome Registry (CATOR) data on 1776 women to show important differences between women under 30 years old and women over 30. Women under 30 were four to six times more likely than older women to use illicit drugs such as marijuana, cocaine, other stimulants, and hallucinogens. While alcohol was the most frequently used substance for women of all ages, women over 30 were more likely than women under 30 to be daily drinkers, to use only alcohol and, when they were chemically dependent, to be daily users of tranquilizers or other prescribed drugs. The women under 30 were also more likely to report the following: family histories of alcohol and drug abuse; family histories of physical or sexual abuse; depression and restlessness; sexual problems; and outpatient psychiatric care.

Other studies have found striking differences among women in socioeconomic factors often thought to predict successful treatment outcomes. Most important among these seems to be insurance coverage, which is strongly related to completion of treatment in both men and women (Amaro, Beckman, and Mays, 1987; Beckman and Bardsley, 1986.)

Women whose alcoholism follows one or more episodes of clinical depression may have different treatment needs from women whose alcoholism develops in the absence of pre-existing depressive episodes (Wilsnack, 1983; Blume, 1986) Shuckit and his colleagues (1969) have distinguished between "primary alcoholism" (alcoholism not associated with a pre-existing psychiatric problem) and "secondary alcoholism" (alcoholism associated with a pre-existing mental disorder—most often, for women, an affective disorder). The treatment implications of these differences are not yet clear, however. Women with affective disorders appear to have alcoholism treatment outcomes that are equal to or even better than those of other women in treatment

(Blume, 1980). Also, there is at this point no clear rule of thumb for distinguishing "primary" alcoholism from "secondary" alcoholism during treatment.

Patient-Treatment Matching

Given the heterogeneity of women in treatment for alcoholism, it is important to look at the question of patient-treatment matching: What treatment strategies are most effective for which women? The National Institute on Alcohol Abuse and Alcoholism (NIAAA) is currently funding a study investigating the possibility that a recent history of depression, panic disorder, or agoraphobia may predispose women to relapse after alcoholism treatment.[1] This study assesses the value of imipramine in treating the symptoms of these disorders and preventing relapse. Another NIAAA-funded study is exploring the feasibility and usefulness of distinguishing between primary and secondary alcoholism in patients who are in treatment for alcohol problems. This study, which examines women, as well as men, will also assess the value of an antidepressant (desipramine) in helping alcoholics with a primary depression and secondary alcoholism to maintain sobriety.[2] A third NIAAA study is investigating the relationship between anxiety and alcoholism to determine whether some women may use alcohol to "self-medicate" for anxiety. The research will also examine the relationship between women's alcohol use and anxiety, on the one hand, and the phases of their menstrual cycle, on the other.[3]

The fact that both depression and prescription drug dependence are common in alcoholic women, however, indicates that caution is needed when using pharmacological therapies. One small study comparing male and female patients referred to an alcoholism clinic found that the women were less likely than the men to see alcohol as a primary problem and more likely to blame their drinking on other difficulties. Women were more likely to report depression and anxiety and to try to obtain prescriptions for psychotropic drugs at the clinic (Thom, 1987). Too great a readiness to prescribe mood-altering drugs for women with alcohol problems is dangerous. Such prescriptions may result in the formation of a new addictive pattern or the exacerbation of an old one. Since many women enter treatment with combined dependency on alcohol and sedative drugs (Blume, 1986). In many cases the anxiety and

---

1. Frederick M. Quitkin, Columbia University: "Imipramine Treatment: Alcoholics with Affective Disorders," R21 AA07688-01.

2. Barbara J. Mason, Cornell University, "Longterm Antidepressant Treatment of Alcoholism," (R23 AA06866-01A1).

3. Daniel R. McLeod, Johns Hopkins University, "The Anxiolytic Effect of Alcohol in Clinically Anxious Women" (R23 AA06713).

depression reported by alcoholic women are related to alcohol use and dependency rather than being symptoms of underlying psychiatric disorders.

Special Needs

Not all issues of concern to women being treated for alcoholism require special treatment modalities. Some may be addressed by existing treatment practices. Women, for example, tend to begin treatment with lower self-esteem than men, but a year after beginning treatment they are similar to men in self-esteem, even without any kind of special programming in treatment (Beckman, 1978). Others may need to be addressed through programs that supplement standard treatment modalities. One potential area for special programming involves sexual issues such as sexual dysfunctions and sexual abuse history. Studies of women in substance abuse treatment indicate that as many as 3/4 of women in treatment may have a history of sexual abuse (Rohsenow, Corbett, and Devine (1988). Many of these women experience sexual dysfunction (Skorina and Kovach, 1986) and may feel incapable of engaging in sexual activity unless drinking or using drugs (Fewell, 1985). Although men and women in treatment both frequently report sexual dysfunction, women manifest more distress about sexual problems when they occur (Heiser and Hartmann, 1987). In a study of 117 women in treatment for alcoholism, Kovach found that 40 percent of those with a history of sexual abuse exhibited symptoms of post-traumatic stress disorder (Kovach, 1986). She argues that delayed stress symptoms can precipitate relapse and should be dealt with in treatment. A number of other researchers also believe that failure to address sexual issues during alcoholism treatment may contribute to early relapse in a significant subset of treated individuals (Rohsenow et al., 1988; Skorina and Kovach, 1986; Fewell, 1985). Because these issues may be difficult for women to discuss in the mixed-sex groups that characterize most treatment facilities, there may be a need for special programming in this area.

RESEARCH NEEDS

In her contribution to an NIAAA monograph on women and alcohol, Sheila Blume listed three important research needs concerning women in treatment (1980). They were:

- Carefully controlled studies comparing different methods of treatment in groups of women that differ with respect to severity of alcoholism, presence or type of psychiatric co-morbidity, age, sex, or socio-economic status.

- A specific study of the long-term treatment of women alcoholics with affective disorders.
- Further studies on sex hormone levels in alcoholic women.

Studies along these lines are now being funded by NIAAA and have been described in this paper. They should yield important findings. Other research needs that have been identified include:

- Studies examining similarities and differences among alcoholic women within various demographic and clinical subgroups (Wilsnack, 1982).
- Research on the concepts of "enabling" and "codependency". These terms denote family patterns that are thought to interfere with recovery. Such research should be free of gender bias and should assess the effectiveness of the spouse and family-oriented interventions used in so many treatment programs (Roman, 1988).
- Studies of the effects of interventions directed toward the offspring of alcoholics that compares their impact on males and females (Roman, 1988).
- Research on patient and therapist expectancies concerning the effectiveness of various treatment modalities for women and on how these expectancies affect women's treatment outcomes (Vannicelli, 1984).

Vannicelli (1984) makes the general recommendation that researchers studying alcoholism treatment take care to specify the precise sex composition of the initial sample studied and of their follow-up sample. She also emphasizes that outcome results should be presented separately for men and women. Blume (1980) goes even further, recommending that all titles of articles on alcoholism treatment specify whether men, women, or both men and women studied. Vannicelli also urges that authors refrain from drawing conclusions about "alcoholics" when there are no women in the sample studied or when the number of women is negligible (1984).

REFERENCES

Amaro, H.; Beckman, L. J., and Mays, V. M. (1987) "A Comparison of Black and White Women Entering Alcoholism Treatment," *Journal of Studies on Alcohol*, 48(3) 220–227.

Annis, H. M. and Liban, C. B. (1980) "Alcoholism in Women," in *Alcohol and Drug Problems in Women*, Kalant, O. J., ed. New York: Plenum Press, 385–422.

Beckman, L. J. and Amaro, H. (1986) "Personal and Social Difficulties Faced by Males and Females Entering Alcoholism Treatment," *Journal of Studies on Alcohol*, 47, 135–145.

Beckman, L. J. and Amaro, H. (1984) "Patterns of WOMEN"S USE OF Alcohol Treatment Agencies," in *Alcohol Problems in Women*, S. C. Wilsnack and L. J. Beckman, eds. NY: Guildford Press.

Beckman, L. J. and Bardsley, P. E. (1986) "Individual Characteristics, Gender Differences and Drop-out from Alcoholism Treatment," *Alcohol and Alcoholism*, 21(2) 213–224.

Blume, S. B. (1986) "Women and Alcohol," *Journal of the American Medical Society*, 256 (11) 1467–1469.

Blume, S. B. (1980) "Researches on Women and Alcohol," *Alcohol and Women*, Rockville, MD: National Institute on Alcohol Abuse and Alcoholism (DHEW Publication No. ADM-80-835).

Fewell, C. H. (1985) "The Integration of Sexuality into Alcoholism Treatment," *Alcoholism Treatment Quarterly*, 2(1) 47–60.

Harrison, P. A. and Belille, C. A. (1987) "Women in Treatment: Beyond the Stereotype," *Journal of Studies on Alcohol*, 48(6) 574–578.

Heiser, K. and Hartmann, U. (1987) "Disorders of Sexual Desire in a Sample of Women Alcoholics," *Drug and Alcohol Dependence* 19 145–157.

Kovach, J. A. (1986) "Incest as a Treatment Issues for Alcoholic Women," *Alcoholism Treatment Quarterly* 3(1) 1–13.

Laign, J. (1987) "How far Have We Really Come Baby? Women's Addiction Treatment in 1987." *Focus on Chemically Dependent Families*. September–October 14–30.

National Institute on Alcohol Abuse and Alcoholism (NIAA) (1987) *Sixth Special Report to the U.S. Congress on Alcohol and Health,* Rockville, MD: DHHS Publication No. (ADM) 87–1519.

National Institute on Alcohol Abuse and Alcoholism (NIAAA), (1986) *A Guide to Planning Alcoholism Treatment Programs*, Rockville, MD: DHHS Publication No. (ADM) 86–1430.

National Institute on Drug Abuse (1989) *Highlights from the National Drug and Alcoholism Treatment Unit Survey (NDATUS), Rockville, MD: NIDA.*

Roman, P. (1988) *Women and Alcohol Use: A Review of the Research Literature.* Rockville, MD: National Institute on Alcohol Abuse and Alcoholism (DHHS Publication No. ADM 88–1574).

Rohsenow, D. J.; Corbett, R., and Devine, D. (1988) "Molested as Children: A Hidden Contribution to Substance Abuse." *Journal of Substance Abuse Treatment*, 5 13–18.

Schuckit, M.; Rimmer, J.; Reich, T.; King, L. J., and Winokur, G. (1969) "Alcoholism I: Two Types of Alcoholism in Women," *Archives of General Psychiatry*, 20 301–306.

Shapiro, S.; Skinner, E. A.; Kessler, L. G., (1984) "Utilization of Health and Mental Health Services: Three Epidemiologic Catchment Area Sites," *Archives of General Psychiatry,* 41 971–978.

Skorina, J. K. and Kovach, J. A. (1986) "Treatment Techniques for Incest-Related Issues in Alcoholic Women," *Alcoholism Treatment Quarterly*, 3(1) 17–30.

Thom, B. (1987) "Sex Differences in Help-seeking for Alcohol Problems: 2. Entry into Treatment," *British Journal of Addiction,* 82 989–997.

Vannicelli, M. (1984) "Treatment Outcome of Alcoholic Women: The State of the Art in Relation to Sex Bias and Expectancy Effects," in *Alcohol Problems in Women*, S. C. Wilsnack and L. J. Beckman, eds. New York: The Guilford Press.

Verbrugge, L. M. and Wingard, D. L. (1986) "Sex Differentials in Health and Mortality," *Health Matrix*, 5(2) 3–19.

Wilsnack, S. C. (1982) "Alcohol Abuse and Alcoholism in Women," *Encyclopedic Handbook of Alcoholism*, NY: Gardner Press.

CHAPTER 12

Opiates

Marsha Rosenbaum, Ph.D.
Sheigla Murphy, M.A.

INTRODUCTION

Women opiate addicts have been maligned since the Harrison Act of 1914, which marked the beginning of the criminalization of the use of narcotics. The words "woman addict" conjure up, for most people, the image of junkie-whore-convict. The field of drug abuse, including research, has until the mid-1970's ignored women. It was assumed that there were too few women addicts to matter. Furthermore, women were not seen as having distinctive characteristics and problems and were therefore lumped together analytically with men.

Paradoxically, Richard Nixon and the women's movement unwittingly joined hands in the early 1970's to change this. Nixon's version of the "war on drugs" called attention to the so-called drug epidemic and pumped millions of dollars into treatment and research. Later, the thousands of addicts who entered the newly-created treatment facilities were counted as proof of the epidemic. Simultaneously, the women's movement of that same era demonstrated that women had drug experiences and problems which, as in other facets of life, were different than those of men. Consequently, by 1976 the National Institute on Drug Abuse (NIDA), having been allocated large amounts of money for research and treatment, became formally interested in women. Among their solicitations were proposals to do ethnographic, "street" studies of women addicts. In those years, the focus was on women, with crime as a concomitant concern. Currently, the focus is on women, drugs and AIDS.

We began work on "The Career of the Woman Addict" in 1977. It was to be an ethnography, including 100 depth interviews with unincarcerated, not-in-treatment women addicts. We advertised and used the snowball, or chain referral, method to locate respondents. We posted notices in high drug-using areas, on any surface that would take a staple. In this way, we were able to do fieldwork simultaneous to publicizing our study, because two women (both pregnant) posting signs reading, "Women Addicts needed to participate in research project. . . ." was not an ordinary occurrence in the community at that time (now, with AIDS outreach, it is commonplace). People wanted to know what we were doing, and we spent a lot of time talking with addicts in the streets while posting our signs. As women, we tried to be sensitive to appropriate locations for reaching women, and found ourselves stapling our notices in laundromats, supermarkets and hospital clinic waiting rooms. It wasn't until the end of our study, when we had theoretically saturated "street" categories, that we interviewed women addicts in treatment programs and in jails.

While studying woman addicts, we learned that treatment was not separate from but an integral part of the heroin world. Addicts routinely went into treatment, just as they went to jail (Rosenbaum and Murphy, 1981). Methadone maintenance in particular had become a phenomenon to be reckoned with, and again women in treatment had special, distinctive problems. NIDA agreed, and awarded us a grant to study women on methadone. For this project, we interviewed 100 women who were currently enrolled in methadone programs. Using a processual career model, we extensively interviewed each of the women.

In this paper, we will begin with a discussion of our major findings about women and heroin, moving on to discuss women on methadone. Next we will look at the outcome of women's addiction. Finally, we will look at current concerns regarding women and drugs, with a focus on AIDS.

HEROIN: A CAREER OF FUNNELING OPTIONS

Women enter the heroin world for a number of reasons, among them the perception that it is an expansion of life options. They are attracted to heroin because they want to be part of a social scene complete with the appearance (and sometimes the reality) of money, action and the euphoric properties of drugs better known as "getting high." Heroin has the reputation of being the "hardest drug," and as such its users take on an outsider or outlaw status which is sometimes appealing to young women. These attractions add a new dimension to women's lives; lives which they often describe as boring, alienating and without money. Occasionally, women get into heroin through a boyfriend or spouse, but we found that this does not occur as often as the literature suggests. Some of the middle class interviewees (a minority among addicts) were initially attempting to alleviate physical pain, often connected with menstruation. They began with legal prescriptions, and in time purchased street drugs and eventually heroin when these prescriptions were withdrawn by physicians.

Initially, women enjoy the beginnings of a heroin career. But real life as a junkie turns out to be different than the initial, or "honeymoon," phase. In the next stage, women become locked in and the illusion of the expansion of options in the beginning stages transforms into a funneling or reduction of life options. Gifts of heroin so common for the (female) newcomer begin to dwindle. Despite the fact that male heroin addicts claim women have it easier because ". . . they are sitting on a millions dollars. . . ." (quotation: respondent), support of their habits and themselves becomes a major issue for most women addicts. To compound the difficulty of support, nothing is predictable, especially procuring drugs. The heroin world is chaotic and finding a consistent source of good street drugs takes up most of the addict's time, so holding down a job is nearly impossible. Legal employment available to unskilled, under-educated women rarely pays enough. Most women turn to illegitimate jobs, with better pay and more flexible hours, to earn a living and to pay for their drugs.

Upon entering the world of illegal work, these women become criminals as well as junkies and their lives become full of risk as well as chaos. Such work is usually sex-role related; prostitution, forgery, shoplifting, sales (dealing). Brushes with or entrance into the criminal justice system escalate the funneling of life options in the conventional world. Now the woman has a record. Involvement in criminal activities locks her into the deviant lifestyle.

Ultimately, most women want out, primarily for reasons that are sex-role related. The first is inundation in the heroin life. Because women have little time for anything but heroin and related pursuits, children are neglected, jobs are lost, and numerous health problems (including frequent withdrawal) are experienced. Second, most women want out when they become pregnant accidentally. They generally want to clean up in order to avoid having a "strung out" baby. Third, at about thirty, many women experience "burnout," coupled with panic, and often ask themselves, "Is this who I want to be forever?" The answer is almost always, "No." Finally, the recognition of funneling options is a major factor in women's desire to get out of the heroin life: losing children (through neglect and/or social service agencies interventions), either not having had children or becoming pregnant while addicted, and cutting ties with the conventional world (Rosenbaum, 1981).

Since the proliferation of heroin treatment in the late 1960's and early 1970's, women (and men) who want out of the heroin world frequently look to treatment programs for help. The largest single treatment modality for heroin addiction in methadone maintenance, and that is where many women have turned. As primary caregivers for children, they find methadone maintenance is often their only option. Therapeutic communities usually do not accept children, and it is only within the last few years that a small number of programs have been opened to meet the needs of chemically dependent women and their children (Reed, 1987).

SURRENDER TO CONTROL: METHADONE

The initial phase on methadone maintenance can be characterized by surrender to control. Methadone replaces the chaos of the heroin world with a structured routine: mandatory clinic attendance, payment of fees, urinalysis, counseling. There is often a sense of relief: the daily job of getting money is gone; women have more time; and best of all, they are not sick every day. For pregnant addicts, methadone provides stability. Some programs even have a special component for pregnant addicts in which they are given support and education about childbirth and child-rearing. There is an initial appreciation of the control exerted by the clinic and its accompanying structure. The unspoken posture is, "Help me."

In the next stage, "stabilization" (sometimes called "addiction"), an enduring relationship to methadone is established. Women have adjusted to methadone as a drug and have arrived at a correct dosage (the effects of heroin are blocked and withdrawal is prevented). They begin to earn take-home doses and begin to take back some control over their lives. In this stage, a dichotomy is apparent. There are the clinic "successes" who break with the deviant world and use methadone to attempt to re-enter conventional life. There are also the clinic "failures," those who continue to see themselves as part of the heroin world (often they have tried to re-enter but failed) and use methadone as a fall-back drug and lifestyle.

Finally, there is the disillusionment phase on methadone, in which surrender to control becomes resentment of control. The objects of resentment are mandatory reporting to the clinic and consequent travel problems, health concerns, parenting problems (e.g., attempting to be a good role model while on methadone), and identity. Identity problems center around being a "half-junkie" and aging. Women complain that they are in a sort of limbo, not a junkie any more but still an addict. If attempting to function in the conventional world, one's status, as a methadone client must be concealed; hence, these women continue to carry the stigma of addiction although they have exited the heroin world. Aging plays an important role in disillusionment with methadone. Women take stock of their lives and wonder if this (methadone) is what they will be doing for the rest of their lives. Most do not want to be on methadone indefinitely (Rosenbaum, 1982).

Obstacles to detoxifying from methadone are numerous. There are few visible formulas for success or role models who have detoxified without returning to drug or alcohol abuse. Successful detachment from methadone means staying away from the clinic. Methadone counselors have told us if they see a detoxified client hanging around the clinic they assume the client is not doing well. Such a client is likely to be either looking for heroin or trying to buy someone's take-home doses. Therefore, for the most part, there are no successful ex-methadone patients around the clinic to act as models. Paradoxically, being a successful methadone patient can impede detachment from methadone. Women who are making it in the conventional world, often holding down a job and raising children, cannot afford to go through an unsuccessful detoxification. This would mean being "sick" for an indeterminate period of time and possibly throwing their lives into chaos. Through methadone they have achieved a delicate balance, and detoxification would threaten the gains made. Additionally, clinic "successes" get reinforcement from staff, and giving up this positive feedback can also be difficult.

Perhaps the most pervasive obstacle to getting off methadone is fear. As one clinic counselor told us, "with detoxing, what you hear is what you get." Descriptions, mostly negative, about detoxification circulate around the clinic. Women talk about long-term withdrawal symptoms, emotionally vulnerability and a lack of social life (clinic attendance and going out for coffee with other clients become a mainstay of some women's social lives) as problematic aspects of attempts to detoxify. An experienced detoxifier told us, "It's like asking you to visit hell for awhile." As a result, only a small percentage of women detoxify from methadone and remain street drug free (Rosenbaum, 1985).

OUTCOMES OF ADDICTION AND TREATMENT

Outcomes of women's addiction and treatment are linked to social class. For those who enter addiction through the working and lower classes, life options are reduced both outside and inside the deviant world. After having been addicted, it is more difficult for women to plug into the conventional world than it is for men. They carry a deeper stigma and as a consequence find it more difficult to secure and keep jobs. Within the deviant world, women's aging makes it more difficult to make it in sex-role-linked hustles such as prostitution. As a consequence, working and lower class women who become and remain heroin addicts end up receiving a pittance of financial assistance *and/or* are enrolled in methadone programs indefinitely, unless new fiscal policies and fee-for-service treatment programs end even this option (Rosenbaum, Murphy and Beck, 1986; Murphy and Rosenbaum, 1987; Rosenbaum, Irwin and Murphy, 1988).

Women who have entered addiction through the middle class (e.g., hippies, medical addicts) have more options if they haven't become too immersed in the deviant world (e.g., too many incarcerations). It is possible for them to readjust because they often possess the resources necessary to carve out a new life. Perhaps most important, they have middle class status and can define addiction as a "phase" rather than a career (see Rosenbaum, M., 1985 for a discussion of factors contributing to successful detoxification from methadone).

For women and their children, the prognosis is grim. The war on drugs has created an unsympathetic posture towards drug use and addiction, and although there is much talk about treatment, in reality there are fewer services for women than there are for men. All this exists in the midst of an AIDS crisis in which women addicts are seen as possible major conduits of the virus through needle-sharing and prostitution. In California short-duration methadone maintenance is still the norm. Thus we are seeking shrinking numbers of publicly funded treatment slots and shorter duration treatment episodes, recidivism to heroin/needle use, and increased risk of AIDS transmission.

CONCLUSIONS

In this era of AIDS, we must utilize all the resources created and ethnographic information gathered in this last decade to intervene in the spread of this virus. In many ways, women intravenous drug users are outreach workers' and treatment personnel's greatest challenge. Issues concerning sex-roles and stigma make women IVDUS the most difficult to reach at-risk population.

Women's roles in the various cultures from which they come affect the success of public efforts to structure effective interventions. In some communities, women seldom hang out on the streets or outside their homes without their husbands or boyfriends. Admission to strangers that a man's wife or girlfriend has a drug problem can be problematic. Women trained in innovative outreach methods are necessary to recruit women addicts into services available to them. Outreach workers must be willing to spend time where *women* congregate: in kitchens, laundromats and playgrounds, for example.

Efforts to recruit women into treatment programs must take into account their role as primary caregivers for children. Providing services for children, such as coupons for well baby examinations, or organizing play groups for mothers and children in neighborhoods with chemically dependent populations, will attract women into participation. And perhaps most importantly, facilities for children must be a component of all treatment programs intended to serve women.

Women as well as men must be allowed to reenter the conventional world without the burden of stigma. Addicts have to assume new identities and/or resume viable old ones in order to make successful transitions (Biernacki, 1986). The old drug-using lifestyle must be replaced by a new conventional lifestyle. The organizing principle behind many women's lifestyles today is their job outside the home. Efforts must be made in the workplace to reduce the stigma attached to women who are ex-addicts and to support their attempts to change. In sum, ex-addict identities must be made attainable for women as well as for men.

REFERENCES

Biernacki, P. (1986). *Pathways from Heroin Addiction: Recovery from Treatment*, Philadelphia, PA: Temple University Press.

Murphy S. and M. Rosenbaum. " Money for Methadone II: Unintended Consequences of Limited Duration Methadone Maintenance," Forthcoming, *Journal of Psychoactive Drugs,* 1988.

Reed, B. April–June, 1987. "Developing Women-sensitive Drug Dependence Treatment Services: Why So Difficult?" *Journal of Psychoactive Drugs* 19(2).

Rosenbaum, M. *Woman on Heroin,* New Brunswick, NJ: Rutgers University Press, 1981.

Rosenbaum, M. (1982). *Surrender to Control: Women on Methadone,* Final Report, Grant #1 R01 DAO2442, Rockville, MD: NIDA.

Rosenbaum, M. (1985). *Getting Off Methadone,* Final Report, Grant #R01 DA-02242, Rockville, Maryland: National Institute on Drug Abuse.

Rosenbaum, M. and S. Murphy. January—March, 1987. "Getting the Treatment: Recycling Women Addicts," *Journal of Psychoactive Drugs* 13(1).

Rosenbaum, M., S. Murphy and J. Beck. January–March, 1987. "Money for Methadone: Preliminary Findings from a Study of Alameda County's New Maintenance Policy," *Journal of Psychoactive Drugs* 19(1).

Rosenbaum, M., J. Irwin and S. Murphy. (1988). "Different Strokes: A Typology of Clients Impacted By a New Methadone Maintenance Policy," Forthcoming, *Contemporary Drug Problems.*

CHAPTER 13

Cocaine

Barbara Lynn Eisenstadt, Ed.D

Clinicians and counselors who work with cocaine abusing women often are frustrated by the sense that there are some elusive factors just beneath the surface that are keeping these women locked inside their addiction. As desperate as these women are to give up their use of cocaine, there is still something about the drug that seems to have a powerful hold on them . . . something they will not let go of.

Many cocaine abusing females exhibit symptoms and behaviors that are common to all substance abusing women, including impairment in their capacity to recognize and regulate their emotions, significant problems of self esteem, and major difficulties in their relationships (Khantzian, 1985). Particularly with cocaine abusing women, there seems to be an inability "to differentiate within themselves various feeling states (e.g. anxiety versus depression)" (Krystal, 1982). In clinical work with cocaine addicts, Khantzian also sees them suffering from "related painful affect states" which are thought to be "especially unique to cocaine addicts as these states interact specifically with the psychotropic effects of cocaine to make them powerfully compelling." (1985).

The intensity of this compulsion is what appears to distinguish cocaine abusing women from other substance abusing women, and what seems to make them prone for particularly high relapse rates. The intent of this paper is to identify some hidden manifestations of this compulsion and to focus on three areas that are critical in the treatment of cocaine abusing women: grief, sexuality and assertion. These areas are often mis-perceived by both clients and clinicians, yet a clear understanding of their dynamics can have a major impact on positive treatment outcomes.

GRIEF

Dealing with grief is an important part of all substance abuse treatment and recovery. Grieving can be in connection with loss of friends, loss of family, loss of self esteem, loss of physical health, loss of spiritual values, loss of trust . . . all resulting in large part from addiction. Additionally, individuals often experience overwhelming guilt in the hurt they have inflicted on those they love. This combination of grief and guilt can be dangerously immobilizing.

The treatment of cocaine abuse is compounded by the intense bonding between female cocaine abusers and cocaine, which is more intense than the bonding characteristics found in other addictions. Cocaine abusing women in particular must be assumed to be literally in love with cocaine. Other substance abusers need their drug and want their drug . . . cocaine abusers are in love with their drug. No other lover has ever been so attentive to their needs, so attuned to their moods, so responsive to their demands or has made them feel so beautiful, so special or so vibrantly alive. By the time most women enter treatment they can no longer get a "decent" high and they believe that they hate cocaine, but in reality they are constantly reminiscing about what they were like when they first started using the drug, and are desperately trying to recapture that first incredible high. The memory of the pain of their addiction is overshadowed by the memory of the pleasure.

Thus, when women who have been addicted to cocaine give up their drug, they are giving up much more than an addiction. In addition to the physical and psychological stress of detoxification, over and above the normal grieving and loss issues they may be having, they are giving up the most intense love affair they have ever known. Recognizing that their drug may have also been their lover is something they must understand if they are truly to appreciate the power that cocaine has over them. Making certain that treatment time is allocated to work through the grieving issues involved with giving up a loved one can greatly enhance the therapeutic process, cutting to the core of much denial and rationalization. Clinicians must understand and acknowledge this dynamic to be able to help their clients effectively accept this loss and work through the natural grieving process. To achieve success in recovery, women cocaine abusers must be educated to the stages of the grieving process, so they can realize that the sudden intensity of many of their emotions is directly related to their grieving over the loss of cocaine.

SEXUALITY

The unspoken sexual concerns that substance abusing women bring with them into treatment can often impede their progress more than anything else. Their guilt, anger and humiliation over what they have done sexually during their addiction is often addressed in treatment programs in a variety of ways. Yet there is an additional dynamic involved in the sexuality of cocaine abusing women that does not seem to be readily identified or dealt with.

It must be acknowledged that numerous and unusual sexual encounters are often the norm for cocaine abusing women at some point during their addiction. Whether this is done to please their partners, to obtain their cocaine, to keep someone from talking, or for whatever reason, this misuse of their sexuality is almost always a part of their addiction. As clinicians, we know that it is important for our female clients to have a chance to work through their guilt and/or anger about these encounters. What is not as obvious is that much of their guilt does not just stem from their shame at what they did, but also comes from a very private embarrassment at how much they may have enjoyed some of these experiences. They are quietly terrified that if they let anyone know about these thoughts, they will be perceived as "sick" or "perverted" . . . "It's all right to feel bad about what I did, but I must really be disgusting and crazy to even think about wanting to do it again".

Part of this has to do with their being in love with the drug and part of it has to do with the fact that sex may have never been as pleasurable or as intense as it was when they were using cocaine. It is not unusual for a woman who has been non-orgasmic all of her life to suddenly become multi-orgasmic, or to discover that when her inhibitions are gone she is capable of engaging in and enjoying sexual activities which would normally repulse her. Under the influence of cocaine, neither of these instances are unusual or abnormal, but they are very foreign and out of the ordinary to the women to whom they occur. And they often happen again and again in connection with the continued use of the drug. A piece of the secret terror that many cocaine abusing women carry with them is the fear that although they want desperately to give up cocaine, they won't be able to deny themselves the sexual pleasure that often went along with its use, and therefore, they really won't be able to stay away from the drug.

If we do not help our clients to bring out this fear and to face it directly, they may spend their entire treatment experience in denial, and leave treatment primed for a speedy relapse. Women only groups, where there is a non-voyeuristic and respectful atmosphere, can be highly beneficial in giving female clients the support they need to bring this fear to the surface. Knowing that what they are feeling is not disgusting or abnormal is liberating, comforting, and empowering. Knowing that with the help of their 12-step programs and their sponsor, they will have strong allies in helping them successfully work through their cravings (of whatever kind), can be a reassuring and very positive treatment experience.

ASSERTION

Assertion training is often thought of as too simplistic to be helpful with cocaine abusing women. Yet basic assertion training can be a powerful treatment modality in working with a variety of real life problems. So much of their behavior has been passive or manipulative; they have so often lost respect for themselves as well as the people they have used or who have used them. Teaching women cocaine abusers basic assertion skills opens up a new world of options, enables them to honestly and directly interact with other people, and gives them a stronger sense of self esteem and a new-found respect for themselves and others. The AA and NA message of avoiding "people, places and things" also takes on a new meaning when backed up by some well practiced suggestions for action. Clients will have specific ways to cope, when, and if, they find themselves in potentially risky situations.

Learning basic assertive skills may save the lives of women cocaine abusers in a very direct manner. Substance abusing women, particularly Afro-American and Hispanic women, are at an increasingly high risk for contracting or spreading the AIDS virus (CDC, 1986), not only because of their own drug abuse, but also due to the fact that cocaine abusing women are very often the sexual partners of I.V. drug abusers. "While intravenous drug users comprise the largest group of women at risk for AIDS, the sexual partners of I.V. drug abusers are the second largest group of women at risk" (Mondanaro, 1987). Teaching our clients that they have the right to refuse to participate in risky sexual encounters may be a life-saving lesson that they will receive nowhere else. Particularly for our women clients, whether it be for their own health or their partner's protection, "just saying no" doesn't work. They need the specific experience of rehearsing *how* to say no without jeopardizing their self-esteem and without insulting their partners. The role rehearsals and role reversals utilized in teaching assertion skills can be powerfully effective in giving them a chance to try out different ways to stand up for their rights without demeaning others or embarrassing themselves. It seems that we have both a moral and an ethical obligation to give them any and all protection we can, and assertion training can be an effective, non-threatening, practical and respectful way to do this.

It is hoped that this paper has provided new interpretations and treatment suggestions for some of the hidden dimensions that are characteristic of cocaine abusing women; hidden dimensions that if not brought out into the open in a supportive therapeutic environment, may keep these women trapped in their abuse. More than ever before, due in part to the ever increasing impact of AIDS, cocaine abusing women are highly vulnerable and at great risk for self destruction. Clinicians and counselors, in addition to providing a substantial 12-step foundation, must continue to develop new and creative treatment strategies in order to give cocaine abusing women the best possible chance for a strong and solid recovery.

REFERENCES

Centers for Disease Control (CDC). (1986) Acquired Immunodefiency Syndrome (AIDS) Among Blacks and Hispanic-United States. *Morbidity and Mortality Weekly Report.* Vol. 35. (42): 655–656.

Crowley, T. J. (1987). *Clinical Issues In Cocaine Abuse.* In Fisher, S.: Raskin, A. and E. H. Uhlenhuth (Eds.). *Cocaine, Biobehavioral Aspects.* New YOrk: Oxford University Press. 193–211.

Des Jarlais, D. C.: Chamberland, M. E. & Yancovitz, S. R. (1984). Heterosexual Partners.: A Large Risk Group for AIDS. *Lancet.* Vol. 2 (8415): 1346–1347.

Dupont, R. L. (1982). Problem of using Cocaine as an Aphrodisiac. *Medical Aspects of Human Sexuality,* 16, p. 14.

Eisenstadt, B. L. (1986). *Women and Cocaine: Leather, Lace, Lies and Pain.* Paper presented at the NYS Division of Substance Abuse Conference, Kiamesha Lake, New York.

Eisenstadt, B. L. (1988). *Women and Cocaine: Breaking The Stereotypes.* Presentation to Second Annual National Conference on Womens Issues. Columbus, Ohio.

Gawin, F. and Kleber, M. (1984). Cocaine Abuse Treatment. *Archives of General Psychiatry.* 41: 903–909.

Kellerman, J. L. (1977). *Grief: A Basic Reaction to Alcoholism.* Center City, Minnesota: Hazelton.

Khantzian, E. J. (1987). *Psychiatric and Psychodynamic Factors in Cocaine Dependence.* In A. Washton and M. Gold, *Cocaine.* New York: Gulliford Press, pp. 229–237.

Khantzian, E. and Khatzian N. (1984). Cocaine Addiction: Is There a Psychological Predisposition? *Psychiatric Annals.* 14: 753–759.

Krystal, H. (1982). Alexithymia and the Effectiveness of Psychoanalytic Treatment. *International Journal of Psychoanalytic Psychotherapy.* 353–378.

Kubler-Ross E. (197–69). *On Death and Dying.* London; Collier-Macmillan Ltd.

Leishman, K. (1986). Heterosexuals and Aids: The Second Stage of the Epidemic. *The Atlantic Magazine.* Feb. 1987: 39–50.

Milkman, H. B. and Shaffer, J. J. (Eds.) (1985). *The Addictions: Multidisciplinary Perspectives and Treatments.* Lexington, Mass; Lexington Books.

Mondanaro, J. (1987). Strategies for AIDS Prevention: Motivating Health Behavior in Drug Dependent Women. *Journal of Psychoactive Drugs.* Vol. 19 (2): 143–149.

Zackon, F., McAuliffe, W. E., and Chien, F. M. N. 1985. *Addict Aftercare: Recovery, Training and Self-Help.* NIDA Treatment Research Monograph, DHHS Publication No. ADM 85-1341. Washington D.C.: U.S. Government Printing Office.

PART F

SPECIAL ISSUES

CHAPTER 14

Black Women
Maxine Womble, M.A.

INTRODUCTION

Despite numerous reports and studies indicating that alcohol abuse is the number one health problem among in Blacks (Harper, 1976; Brisbane and Womble, 1985; Department of Health and Human Services 1985; National Institute on Alcohol Abuse and Alcoholism, 1987), there remains a paucity of empirical research data, the absence of comprehensive epidemiological studies on Black drinking patterns, to say nothing about effective treatment and prevention approaches. The lack of research regarding alcohol and Blacks in general, is even more startling for Black women. A National study conducted in 1984 (Herd and Caetano, 1987) on drinking patterns and problems among Blacks provides us with the most indepth analysis to date.

This paper provides an extensive review of the literature to examine what is known about the use and abuse of alcohol among Black women. The review provides some information on alcohol prevalence and drinking practices among Black women, an overview of some treatment issues is provided with the aim toward achieving more effective treatment for Black female alcoholics. A brief review of education and prevention approaches targeted to Black women is provided.

PREVALENCE OF ALCOHOL AMONG BLACK WOMEN

In an extensive review of race and sex specific studies by Gary and Gary (1985), showed Black women as having higher percentage of abstainers from drinking alcohol, while on the other end of the spectrum there were higher rates of heavier drinkers. In addition, there was a relatively high percentage of heavy escape drinkers (Bailey et al., 1965; Bahr and Garrett, 1976; Kane, 1981; Miller, et al. 1980). These findings were very similar to those in the 1964–65 National survey of drinking practices which included 200 Black respondents. Cahalan and Cisin (1968) also concluded that Black women differed from White women in their higher proportion of abstainers (51% versus 39%) and their higher rate of heavy drinkers (11% versus 7%) according to Herd (DHHS, 1986).

In reinterviews with the sub-sample from the 1964 national study in 1967, Cahalan found that Blacks, along with those of Caribbean and Latin ancestry had the highest rates of social-consequence drinking problems. National surveys in 1979 and 1983 of drinking practices had similar findings in terms of rates of abstainers among Black women (49%) compared to White women (39%) and heavy consumption for Black women (7%) versus 4% for White women (Herd, 1987). In a 1981 National survey of women's drinking problems (Wilsnack et al, 1984) the findings indicated that Black women and White women were similar in reported problems or symptoms due to drinking, but Black women were significantly more likely than Whites to report having six or more drinks in a day four or more time in the preceding year. It should be noted that most researchers attributed high rates of alcohol use and abuse to a permissive culture for female drinking, and to a large percentage of Black women being heads of households (Herd, 1984). Harper (1976) attributed heavy drinking to "psychological reasons of personal misfortune, liveliness, lack of self confidence, or lack of hope and meaning in future living". (p. 36)

Previous research findings differed significantly from the study of Black Drinking practices in Northern California by Caetano and Herd (1984). This study with 1,206 Black respondents most significant representation of Blacks in an alcohol study to date. Alcohol consumption among Blacks was found to be relatively low between the ages 18 and 29, but showed a tremendous increase among persons in their 30's. Females represented 29% of the abstainers with only 6% of the women classified as frequent heavy drinkers. The data also "suggested that drinking patterns among Blacks are influenced more by internal norms originated from common cultural socio-political characteristics than from norms associated with class affiliation in the larger society". (p. 571)

In a National survey of Drinking Patterns Among White, Black and Hispanic Women in the United States, Herd and Caetano, (1984) found that Black women did not have heavier drinking patterns than Whites as reported in other studies, but heavier drinking was more typical among white women. In looking at the predictors of drinking Black women in higher income levels and Black women who are 18–39 are more likely to become drinkers and have greater chances of being heavier drinkers, Black women were found to develop their highest rates of problems, however, at 40 and over. It was also observed that Black women drank primarily in small quantities even if they drank frequently. These findings by Herd and Caetano differ from previous studies. The same differences are attributed to the location of the population surveyed and religion. A significant number of the respondents were from the south and midwest regions of the country where women traditionally have lower drinking rates.

While the research by Herd provides more insight into Black drinking practices, much more is required if we are to have a full understanding of treatment and prevention needs of Black women. Benjamin and Benjamin (1981) expressed the need for research that would take into consideration sociocultural factors to better understand the affect of alcohol on human behavior is to be understood. They also recommended "intraracial studies that address age, regional and occupational differences, and socio-cultural consequences (legal, employment, family problems, etc.) of alcohol use and misuse". (p. 243)

TREATMENT ISSUES AND CONCERNS

There is an absence of research on cultural factors as they relate to effective treatment modalities and treatment outcomes for Black women (Harper, 1976; Dawkins and Harper, 1983). Some authorities suggest that utilizing methods of treatment that continue to dismiss culture as a variable, increases stress factors, and promotes the exclusion of those in need of services (Watts and Wright, 1983). Research is also needed to help understand sex and ethnic differences in the antecedents and consequences of alcoholism and the resulting differences in treatment needs of female alcoholics of various ethnic backgrounds (Gary and Gary, 1985).

In King's (1982) review of studies on treatment, he found few that examined cultural factors, and none that focused exclusively with Black alcoholism. Bentley (1978), focused on the success of treatment as a function of income, age, sex education and martial status. Studies that focused on follow-up and prognosis (Moos and Bliss, 1978; Welte et al., 1978) reported that Blacks had a lower follow-up rate and a poorer prognosis than other groups. Other studies reported Black women were more likely to succeed in treatment. Strayer's (1961) study indicated that Black females showed even stronger motivation for sobriety than Black males, while some studies (Blum and Blum, 1972) suggested that Blacks showed a greater motivation to seek treatment when alcoholism treatment programs recruit and employ outreach techniques. Benjamin (1976) found that among rural Blacks, the physical effects of heavy drinking were reasons most sought treatment, and that rural Black families tended to treat or deal with their alcoholic family member at home. It is therefore important that treatment be developed for women keep in mind the regional, cultural and socioeconomic differences.

Gary and Gary's (1985) review of studies (Carroll, et al., 1982; Corrigan and Anderson, 1982; Miller, et al., 1980) to determine the extent to Black and White female alcoholics in treatment programs found that Black women alcoholics are a very heterogeneous group and more attention must be given to the differences in developing treatment strategies. Miller and his Colleagues (1980) concluded that treatment programs should allow for those distinct drinking patterns and life styles by providing a wide variety of outpatient aftercare options.

Treatment programs for Black women can become more effective, according to Gary and Gary (1985), if consideration is given to the following recommendations: (1) practitioners should understand that the doctrine of colorblindness has outlived its usefulness; (2) clinicians should become more sensitive to the social and cultural contexts in which Black women have lived and continue to live in this society; (3) alcohol therapists should try to broaden the treatment of patients to include relevant family members and kins; (4) building a social support network through the involvement of friends, neighbors, and alumni of treatment programs should be an integral part of a treatment plan; (5) the Black church should be seen as an important source of support and resources for both the patients and the practitioners; (6) promoting racial consciousness or pride should receive greater priority in the implementation of a treatment strategy of Black female alcoholics; (7) negative self-image and low self-concept need to be actively and directly addressed in the treatment of those alcoholic clients who have these traits; (8) confrontation and supportive relationship building are important process techniques for responding to selective personality attributes of Black female alcoholics; (9) therapists and counselors should develop for their patients self-development programs which emphasize coping and life management skills and education regarding alcoholism; (10) feminist consciousness raising should be integrated into the therapeutic process; (11) more consideration should be given in treatment planning to assigning women therapists, especially Black females, to Black women alcoholic; and (12) treatment staff members, should develop a better system for monitoring the physical health of alcoholic clients.

These are excellent guides for developing culturally relevant programs for Black women. In keeping with these recommendations it is suggested that treatment programs for Black women should be mindful of the following concerns.

Staff Considerations

There are differences of opinion about whether therapists should be of the same race and sex (gender) for treatment to be successful. Some authorities (Higgins and Warner, 1973; Vontress, 1971) suggest that race of the helping person is not a significant factor in therapy. While others (Beverly, 1975; Smith, 1973) maintain that the helping person must be of the same racial identity or have an understanding of the background and behavior of the patient. Murphy (1977) pointed out that one problem related to transracial relationships stem from the basic difficulties in developing relationships of trust which makes genuine self-disclosure possible. Self-disclosure is regarded by some counseling therapists as basic to any kind of successful therapeutic intervention.

In a study by Woken, et al., that compared the attitudes of Black middle class and lower class college students toward psychotherapy with attitudes of White middle class students, they found that cognitively, middle class Blacks are favorably disposed toward psychotherapy and often seek it. They, however, share a general racial distrust of situations in which they are asked to explore the personal dimensions of their lives. This distrust is exacerbated in transracial therapy situations. Less affluent Blacks have a negative attitude toward psychotherapy mode of intervention, and do not seek out this kind of treatment (Murphy, 1977).

If the therapist views Black clients through colorblind eyes, without regard for their race, culture, socio-economic status, and the American political context in which they live, the possibility of effective treatment is unlikely (Brisbane and Womble, 1985). Black women suffer from alcoholism in ways that are unique to them because they are Black, female and alcoholic, and are, therefore, in a triple bind. Frequently they experience problems that are different from those experienced by other ethnic and racial groups, both in their quest for treatment and in their ability to maintain sobriety.

Religion and Spirituality

Treatment programs and staff need to understand the value of religious beliefs and spirituality in the lives of most Blacks. The church provides an emotional release of tensions accumulated through experiences faced in an oppressive society. The singing of spiritual preaching, and group interaction all help to promote group solidarity and give them a sense of identity (Staples, 1976; Hill, 1972; Knox, 1985). Knox indicated that an assessment that does not factor in spirituality may work against a positive treatment outcome. Spirituality, according to Knox, centers on the beliefs that people can make contact with a superior power and call upon a higher power for solutions to their problems. In this process they gain strength toward solving interpersonal, physical, and economic problems (Brisbane and Womble, 1985).

Research by Herd (1984) found that females without religious affiliation formed the heaviest drinking group. Black females belonging to fundamentalist religions like the Church of God, Assembly of God,

Seventh Day Adventist, Jehovah's Witness, Sanctified and Pentecostal, had twice as many abstainers as Catholics and Baptists. These findings regarding the importance of religion to Blacks were consistent with those of Gary and Gary. Lowen Fish (1977) research also suggested that Black women who were churchgoers were less likely to drink. Thus, therapist need to understand the importance of religious involvement of Black women as part of the treatment process.

Black Family Involvement

Family members should be an integral part of the treatment process for Black female alcoholics (Harper, 1979). some authorities believe that because Black women assume considerable responsibilities in providing and caring for the family, her alcoholism produces a great deal of guilt (Davis, 1979; Harper, 1979). These guilt feelings could interfere with her desire to seek treatment. Since Blacks tend to look upon seeking treatment as a sign of weakness and not strength, the guilt ridden Black female will try to keep her alcoholism secret from the family. Therefore, special culturally sensitive outreach efforts must be made to encourage her to seek treatment.

Attention should also be given to engage the family members in the treatment process which will be helpful in the recovery process. Treatment staff should be knowledgeable about "the Black Family" which should be seen as a multi-generational group. Blacks tend to informally adopt relatives who are close family friends. They become cousins, aunts, and uncles without being related by blood. These adopted relatives become primary part and part of the extended family grouping (Chunn, 1983). As extended family members they become part of a family's central support system, and may exert a great deal of influence.They can sometimes be counted on to help in ways that blood relatives may find too painful. Parents might turn to their daughter's godparents, for example, to help her with her daughter's drinking problem because it is too painful for the parents to handle.

Treatment for Black women should be holistic. There are numerous psychological, physical, social and medical problems that must be addressed. If treatment is to be effective, practical approaches to the problems is a must. Treatment can not be easily achieved while the client is worried about being homeless, in need of financial support or child care services. Programs must, also, be designed to meet some of the unique needs of women clients such as employment, education, and training. However, unless the dimensions of race, gender, culture and socio-political factors are included in the assessment and treatment phases (Brisbane, 1985) sobriety will be difficult to maintain.

PREVENTION APPROACHES

Targeting groups for education, prevention and intervention who are at high risk of premature death from alcohol misuse may provide the greatest benefit in reducing alcohol-associated premature mortality in the United States (Center for Disease Control, 1986). Since Blacks, with the exception of youth, are overrepresented in most indirect measures of alcohol problems, attention must be given to prevention strategies, specifically targeted to them, and should be a National priority.

In his review of the literature King (1982) could not find any studies that focused on prevention approaches for Black women. He found that most of the studies on prevention centered on the examination of socioeconomic strategies for control and prevention. The questions that is frequently raised is whether the abundance of liquor outlets in urban Black communities contribute to the alcoholism rate. The consumption model argues for strong effects of outlet availability on per capita consumption and alcohol rates. Parker and Harman, (1978) found that alcohol availability, independent of other factors, appears to influence alcoholism rates. In addition, they found that per capita income appears to influence consumption rates but not rates of alcoholism (King, 1982). While more definitive research in this area is necessary, Black communities need to be more involved in the control of liquor outlets given the implications for prevention.

Alcoholism prevention advocates need to gain the cooperation of Black organizations and institutions to participate in the development and dissemination of culturally relevant alcohol education and prevention information. The clergy, being the most influential leader in Black communities, should be empowered with information which can be filtered to the religious institutions. These efforts should target professional groups and social organizations with large female memberships.

SUMMARY

Based on the most recent research, Black women seem to have higher rates of abstainers and fewer heavy drinkers, and experience more alcoholism problems at 40 years of age and over. Religion and age seem to have greater influence on drinking patterns than do marital status and education attainment. Black women with higher socio-economic status, and those unmarried and younger are more likely to be drinkers. Treatment programs for Black female alcoholics should be holistic in their approach, employ staff who are not colorblind and are knowledgeable about Black culture, Prevention efforts should include key community based Black leaders and target organizations with large female membership.

REFERENCES

Benjamin, R. and Benjamin M. (1981). Sociological Correlates of black driving: Implications for research and treatment. *Journal of Studies on Alcohol.* (Suppl. 9): 241–245.

Benjamin, R., Rural black folk and alcohol. In : Harper, F. D. Ed.: Alcohol Abuse and Black American. Virginia: Douglass Publisher, Inc. p. 50–60.

Bentley, J. L., The relationship of Psychometric and demographic variables to the success of alcoholics in treatment. In: King, L. M. Alcoholism: Studies Regarding Black Americans.

Beverly, C. (1983). Toward a model for counseling the Black alcoholic Journal of nonwhite concerns in personnel and guidance. In: Watts, T. D. and Wright, R., eds., *Black Alcoholism.* Illinois: Charles C. Thomas. p. 73.

Blum, E. and Blum, R. (1983). Alcoholism: Modern Psychological Approaches to Treatment. In: Watts, T. D. and Wright, R., eds. *Black Alcoholism.* Illinois: Charles C. Thomas. p. 75.

Caetano, R., and Herd, D. (1984). Black Drinking Practices in Northern California. *American Journal of Drug & Alcohol Abuse,* 10(4) 571–587.

Chunn, J. (1983). The Evolution and Essence of the Black Family: Treatment Implications, Bulletin of the N.Y.S. Chapter of the NBAC, Vol. 2 No. 2.

Davis, F. T. (1976). Counseling the Black Alcoholic. In: Harper, F. D., ed. Alcohol Abuse in Black in Black America. Virginia: Douglass Publishers, Inc. p. 85.

Dawkins, P. D. and Harper, F. D. (1983). Alcoholism Among Women: A Comparison of Black and White Problems Drinkers. The International Journal of the Addictions, 18(3), 333–349).

Gary, L. E. and Gary, R. B. (1985). Treatment Needs of Black Alcoholic Women. In: Brisbane, F. L. and Womble, M., eds. *Treatment of Black Alcoholic,* New York: The Haworth Press, p. 97–114.

Harper, F. D. (1976). *Alcohol Abuse And Black America,* Virginia: Douglass Publishers, Inc.

Herd, D. and Caetano, R. (1987). Drinking Patterns and Problems Among White, Black and Hispanic Women in the U.S.: Results from a National Survey. San Francisco: Alcohol Research Group, Medical Research Institute.

King, L. M. M. (1982). Alcoholism: Studies Regarding Black Americans. Alcohol and Health Monograph 4: Special Population Issues DHHS Publication No. (ADM) 82–1193, Washington, DC: U.S. Govt. Print Office, pp. 385–407.

Knox, D. H. (1985). Spirituality: A tool in the Assessment and Treatment of Black Alcoholics and Their Families, In: Brisbane, F. L. and Womble, M., Treatment of Black Alcoholics, New York: The Haworth Press, p. 31.

Lowenfish, S. K. (1983). Women Alcoholic: Her Clinical and Social Emergence. In: Watts, T. D. and Wright R., eds. Black Alcoholism. Illinois: Charles C. Thomas, p. 47.

Miller, K. D., et al. (1985). Differences in demographic characteristics, drinking history and response to treatment of Black and White Women seen at an alcohol detoxification center. In: Brisbane, F. L. and Womble, M., eds. *Treatment of Black Alcoholics,* New York: The Haworth Press, p. 97–114.

Murphy, J. (1977). "Mental Health in a cultural context: an examination of some of the limitations of Traditional Approaches to Therapeutic intervention with Black People. Paper presented at workshop on Minority Utilization. Atlanta, Georgia.

Moos, R. and Bliss, F. (1978). Difficulty of follow-up and outcome of alcoholism treatment. J. Stud Alcohol, 39(3) 473–490. Alcohol and Health Monograph 4: Special Population Issues. DHHS Publication No (ADM) 82–1193, Washington, DC: U.S. Govt. Printing.

National Institute on Alcohol Abuse and Alcoholism. (1987). Sixth Special Report To The U.S. Congress on Alcohol and Health. Rockville, Maryland.

Parker, D. A. and Harman, M. S. (1982). The distribution of consumption model of prevention of alcoholic problems: A critical assessment. In: Alcohol and Health Monograph 4—Special Population Issues. DHHS Publication No. (ADM) 82–1193, Washington, DC: US. Govt. Print Office, p. 385–407.

Smith, E. (1983). Counseling the Culturally Different Black Youth. In: Watts, T. D. and Wright, R., eds. *Black Alcoholism*. Illinois: Charles C. Thomas, p. 73.

Strayer, R. (1983). A study of the Negro alcoholic. Quarterly Journal of Studies on Alcohol: In: Watts, T. D. and Wright, R., eds. Black Alcoholism, Illinois: Charles C. Thomas, p. 75.

Watts, T. D. and Wright, R. (1983). Black Alcoholism. Illinois: Charles C. Thomas.

Welte, J., et al. (1978). "Predictors of follow-up completion and abstinence in an Alcohol Rehabilitation Outcome Study". Paper presented at NCA meeting, St. Louis.

US Department of Health and Human Services. (1986). Report of the Secretary's Task Force on Black and Minority Health. Vol. VII., Washington, DC: US Government Printing Office.

CHAPTER 15

Mexican-American Women
Juana Mora, Ph.D.

Mexican-American women are members of one of the fastest growing ethnic group in the U.S. Recently released data from the U.S. Bureau of the Census indicate that between 1980 and 1987 the U.S. Latino population increased from 14.5 million to 18.8 million representing a 30% increase compared to a 6% increase among non-Latinos during the same period (U.S. Bureau of the Census, Advance Report, 1987). In 1985, there were 8.5 million Latina women in the U.S., the majority over-represented in the lower end of the socio-economic scale.

MEXICAN-AMERICANS AND ALCOHOL USE

The current and projected growth rate of the Mexican-American population is magnified by well documented indications that Mexican-Americans suffer disproportionately from problems of poverty, low educational attainment, under and unemployment, low adult literacy and lack of health care (U.S. Bureau of the Census, 1987), factors associated and exacerbated by alcohol and other drug use. Moreover, Mexican-Americans may be at an elevated risk for alcohol and other drug use because of their concentration in large metropolitan urban centers and inner cities, known for a higher prevalence of alcohol and drug use and associated conditions (HHS Secretary's Task Force Report VII, 1986). As the Mexican-American population grows, primarily through natural increase, it is expected that in the next 20 years a very large proportion of the total U.S. youth population will be of Mexican-American ancestry (Santiestevan & Santiestevan, 1984). For young Mexican-Americans alcohol and other drugs are implicated in many of the social problems among this age group, including teenage pregnancy, school drop-out and delinquent behavior. Mexican-American women are also a high risk group for alcohol-related problems. They are over-represented among those living below the poverty level and among those having children at a young age. These conditions combined with issues of language differences, sub-standard and over-crowded housing, immigration status and discrimination compound the stresses caused by alcohol abuse within the family.

MEXICAN-AMERICAN WOMEN AND ALCOHOL USE

The research literature on drinking practices and problems among Mexican-American women is not extensive. The few national and local surveys of alcohol use that specifically examine drinking patterns within this group, consistently describe this segment of the population as primarily non-drinkers or occasional drinkers (Alcocer, 1979; Holck, et al., 1984; Caetano, 1985). The lack of research emphasis on women is partially due to a disagreement in the alcohol field that alcohol is in fact a problem for Mexican-American women. In contrast to the excessive and problematic use of alcohol among Mexican-American men, survey research has not shown a similar degree of drinking among Mexican-American women. However, there is increasing evidence that drinking practices and norms are changing among Mexican-American women and that the extent of alcohol use may be more prevalent than survey research currently indicates.

OVERVIEW: DRINKING PATTERNS

In self-report surveys, Mexican-American women have consistently reported that they either do not drink or drink occasionally. Raul Caetano (1985), for example, in the first national Hispanic alcohol survey found that nearly half or 47% of the women identified themselves as "abstainers" (in this study defined as someone who drinks less than once a year or has never drunk alcoholic beverages). Another 24% of that same sample reported that they drank less than once a month. In a 1979 survey of low-income Mexican-American women in Texas, Maril and Zavaleta reported that 86% of the women in their sample had not consumed alcohol at any time during the last year. In a local study of alcohol use among 200 Mexican-American couples in San Jose, California, 66% of the women identified themselves as abstainers and 33% as infrequent drinkers (Corbett, Mora & Ames, forthcoming). Two other studies (Alcocer, 1979; Holck, et al., 1984) have reported similarly high abstention rates for Mexican-American women.

This self-report data suggests that Mexican-American women are primarily non-drinkers or light drinkers. However, researchers and health care professionals recommend caution in the interpretation of the data since there may be serious underreporting of actual use. It is possible given the cultural sanctions against female drinking, that Mexican-American women are reluctant to report, even with proper assurances of confidentiality, whether they are drinking, how much they are drinking and what problems they may be experiencing as a result of their drinking. While a large proportion of the Mexican-American female population, especially immigrant women, are reporting very little drinking, what is less clear from the data is who are the women who are drinking and what kinds of problems are they encountering as a result of their drinking.

Maril and Zavaleta in the 1979 survey of low-income Mexican-American women found that a small number of their female sample did report some drinking. These women were the younger women and middle-aged married women with higher levels of education. Similarly, by examining age, education, income and generational status variables, Gilbert (1987) found in a sample of Mexican-American women in California that it is primarily the immigrant women who are reporting high rates of abstention or light drinking and the younger U.S. born women who are reporting moderate and heavier drinking. Raul Caetano (1985) also found in his national survey that drinking levels increased among women with higher levels of income and education. It seems that what is taking place is that as Mexican-American women are entering new social and work arenas, they are participating in social events and activities where drinking takes place and as they do this, they are shifting away from lighter drinking towards heavier "high-risk" drinking. Some researchers have suggested that changing social expectations and family roles for these women is adding stress and contributing to the higher drinking levels. Although the number of women reporting this type of drinking is small compared to the light or non-drinkers, a research emphasis on this population would provide specific information on problems and issues related to alcohol use within this group.

There is one other group that has not been studied but has been identified by service professionals as a high risk group in need of alcohol services. These are Mexican-American women married to or living with a family member who is drinking problematically. My experience in working with these women and the experience of others in the alcohol service field is that these women perhaps suffer greater negative health and emotional consequences as a result of living with a problem drinker than other members of the family. These women at times may also be at risk for developing problematic drinking behaviors themselves as a way of coping with the situation.

Developing Alcohol Services

Alcohol becomes a problem for Mexican-American women when they develop problem drinking behaviors or are married to, living with, or related to someone who drinks excessively. Because of the traditional female role of "caretaker" and central figure in the family, the responsibility of maintaining a family under combined poor, unsafe, and alcoholic conditions is a tremendous psychological and physical burden. Immigrant women may not have the social and family networks that are often helpful and most likely will lack the financial and language resources necessary for seeking help. Second generation, U.S. born women may encounter other difficulties in seeking assistance with their own or anothers' alcohol problem—language will most likely not be a barrier, but financial problems and other cultural barriers may still apply.

Mexican-American women, especially immigrant women, will be reluctant to ask sources external to the family for help—if they do seek help it is more likely that they will speak to a trusted neighbor, family member, a "comadre," or a priest. Those internal sources may be helpful in some cases but may not be able to help in chronic situations or for instances when women have a drinking problem of their own. When some women are ready to seek help outside of the family or social network, they may not know where to go or what resources are available in the surrounding community. Too often, there are few options for confidential, affordable, accessible and culturally compatible alcohol services for Mexican-American women.

In order to be effectively utilized, a program must be committed to serving Mexican-American women at all program levels. For a program to reproduce a safe, family-like environment conducive to reaching these women it must adopt a philosophy that will support on-going and specific outreach efforts, home visits, family involvement and a commitment to hiring bilingual, bicultural, recovering Mexican-American women. The development of these services depends on support from the Board of Directors and adoption of language in the by-laws a defining the program philosophy, services, facility planning and staff appropriate to these services.

Effective service programs in Mexican-American communities are woven into the life of the community and become important as they slowly build credibility in that community. Continuity of staff over time is an important element in developing credibility. Another important element in establishing community support is networking with important community institutions such as the local churches, Headstart programs and health clinics. Developing these networks also serves to establish strong and on-going cross-referral links with other service agencies.

Other program elements important to the development of culturally appropriate services for Mexican-American women are, the availability of childcare and a "children are welcome" attitude at the facility, the free use of Spanish, the availability of Spanish AA and Al-Anon meet-

free use of Spanish, the availability of Spanish AA and Al-Anon meetings at the facility and the community and continuous community education and involvement.

Support groups for Mexican-American women are important for the development of positive coping skills and self-empowerment. These are safe environments where these women, perhaps for the first time, learn to trust someone outside the family and find out that they are not alone in experiencing these problems. They learn to cultivate trusting and mutually supportive relationships in a cultural framework that is familiar and acceptable to them.

CONCLUSIONS

Survey research has presented an interesting variation in alcohol use by Mexican-American women based on age, income, and education differences. Specific information on the socio-cultural influences on this variation can be better understood if studied through ethnographic and other qualitative research approaches. Other aspects of alcohol use and problems within this group also need further study, including cultural sanctions and restrictions on alcohol use and help seeking behavior. Further study on alcohol use among younger women, especially those in childbearing years, is also a research priority given the increase and young character of this population.

REFERENCES

Alcocer, A. (1979). Quantitative Study. In technical Systems Institute, "Drinking Practices and Alcohol-Related Problems of Spanish speaking Persons in California." Sacramento, Ca: California Office of Alcohol and Drug Problems.

Caetano, R. (1985). "Drinking Patterns and Alcohol Problems in a national Sample of U.S. Hispanics. In: NIAAA Monograph, *The Epidemiology of Alcohol Use and Abuse Among U.S. Minorities.*

Corbett, K., Mora J., and Ames, G. (in press) "Drinking Patterns and Drinking-Related Problems of Mexican-American Husbands and Wives." Prevention Research Center, Berkeley, Ca.

Gilbert, J. (1987). "Alcohol Consumption Patterns in Immigrant and Later Generation Mexican American Women." *Hispanic Journal of Behavioral Sciences,* Vol. 9, No. 3.

U.S. Department of Health and Human Services. (1986). Report of the Secretary's Task Force: Black and Minority Health, Vol. VII, Chemical Dependency & Diabetes.

Holck, S. E., Warren C., Smith, J. and Rochat, R. (1984). "Alcohol Consumption among Mexican-American and Anglo Women: Results of a survey along the U.S. Mexico Border. *Journal of Studies on Alcohol,* 45: 149–153.

Maril, R. & Zavaleta, N. (1979) "Drinking Patterns of Low-Income Mexican American Women. *Journal of Studies on Alcohol,* Vol. 40, No. 5.

Santiestevan, H. and Santiestevan, S. (Eds.) (1984). *The Hispanic Almanac.* Washington, D.C.: The Hispanic Policy Development Project.

U.S. Bureau of the Census, Advance Report. (1987). "The Hispanic Population in the U.S." Series P-20, No. 416. Washington, D.C.: U.S. Government Printing Office.

CHAPTER 16

Native American Women
Candace M. Fleming, Ph.D.
Spero M. Manson, Ph.D.

Although more than a decade and half old, three community-based epidemiologic studies of adult American Indians and Alaska Natives provide consistent evidence of high rates of alcohol abuse and dependence in this population (Roy, Chaudri, & Irvine, 1970; Sampath, 1974). Shore et al. (1973), for example, interviewed half of the adult members of a Pacific Northwest reservation community and reported that 27 percent of the total population qualified for diagnoses of alcoholism. As in the other two studies, males comprised the greatest percentage of alcoholics, which is congruent with clinical impression and most service utilization data. However, such observations often obscure the fact that many Indian females also suffer from alcohol abuse/dependence and may be at equal or even greater risk for alcoholism than men at various points in their lives.

Despite the fact that Indian females are less likely to drink than the average for the general U.S. population, they clearly are no strangers to alcohol. For example, Moss (1979), in a survey of twenty Indian communities, reported that women comprised 43.2% of her sample of drinkers (n = 1,182). Whittaker's (1962) initial work on the Standing Rock reservation highlighted a patter of increasing consumption of alcohol among Lakota women, and projected a quadrupling of female in comparison to male drinkers over recent generations. This estimate is consistent with a number of subsequent reports specific to this community (Maynard, 1969; Medicine, 1969, 1982; Whittaker, 1982). Indeed, Weibel-Orlando, Long, and Weisner (1984) have found that urban and rural Sioux women now drink as much and as frequently as their male counterparts. This was not true for the other tribal groups included in their study, notably the Navajo and Five Civilized Tribes of eastern Oklahoma, and underscores the significance of intertribal variation. Yet, the proportion of female drinkers appears to be growing in these communities as well (May, 1989).

Such sharp increases stem in part from the rapidly escalating use of alcohol among Indian youth under the age of 20, a group which constituted 44 percent of the Indian population as noted in the 1980 census (U.S. Bureau of Census, 1984). According to Oetting and Beauvais (1985), by the 12th grade, the lifetime prevalence of alcohol use is 96 percent for Indian males and 92 percent for Indian females, which is comparable to the national picture (National Institute on Alcohol Abuse and Alcoholism, 1987). However, Indian youth, girls as well as boys, become involved with alcohol at an earlier age, consume alcohol more frequently and in greater quantities, and suffer greater negative consequences than non-Indian youth.

Recently emerging diagnostic data offers additional insight into the dynamics of heavy drinking as well as alcoholism among Indian women. Manson, Shore, Bloom, Keepers, and Neligh (1987) and Manson, Shore, Baron, Ackerson, and Neligh (in press) have reported a multistage diagnostic study of adult tribal members of three reservation communities drawn from the Pueblo, Plains, and Plateau culture areas within a known cases/non-cases matched control design. Employing a modified version of the NIMH Diagnostic Interview Schedule (DIS), these investigators interviewed 144 Indian women and 54 Indian men about a range of psychiatric disorders, including alcohol abuse/dependence. In their study, the proportion of males who drink heavily as opposed to those who drink moderately was almost twice that of females, as is commonly observed in the general population. Likewise, for both sexes, there was a fairly clear relationship between fewer years of formal education, lower occupational status, early age of first intoxication, and a higher preponderance of heavy drinkers. Yet, when Manson et al. controlled for heavy drinking—and, thus, the effects of potentially different social factors on drinking styles—the ratio of alcoholics to non-alcoholics was approximately the same for Indian men and Indian women (1.15:1), suggesting that the latter run an almost equal risk of alcoholism. Among those meeting criteria for alcoholism, most admitted to blackouts, binges, driving while intoxicated, subsequent car accidents, physical fights, and alcohol-related arrests, with the last two symptoms proving to be more frequent for Indian males and females. Health problems due to drinking were acknowledged by fully a third of these individuals.

Alcoholism, in fact, figures prominently in morbidity and mortality among Indian women. It represents the fifth most frequent cause of death in this segment of the population. Age-specific alcoholism death rates (1983–85) (U.S. Department of Health and Human Services, 1988) are significantly greater at all ages for Indian females than for either White or females from other racial groups (see Table 1).

TABLE 1. Female Alcoholism Mortality Rates Per 100,000 by Age and Race 1983–1985.

	Indian and Native	U.S. All Races	U.S. Other than White Race
Under 5 years	0.0	0.0	0.0
5–14 years	0.0	0.0	0.0
15–24 years	0.6	0.1	0.2
25–34 years	16.6	1.3	4.7
35–44 years	42.4	4.3	12.5
45–54 years	48.3	8.4	14.9
55–64 years	41.7	10.7	18.9
65–74 years	39.3	8.7	12.5
75–84 years	14.4	3.8	4.4
85 years +	9.5	1.0	1.3

Alcoholism also contributes to at least four of the other ten leading causes of death among Indian women: accidents (38.7/100,00), diabetes mellitus (18.2/100,000), kidney or renal disease (7.0/100,000), and homicide (6.7/100,000). Again, each occurs with greater frequency among Indian than non-Indian females.

Unfortunately, the aftermath of chronically heavy drinking among Indian women touches the lives of many of their children as well, contributing to generally dysfunctional family environments which in turn, lend themselves to various forms of abuse and neglect. For example, Fetal Alcohol Syndrome (FAS) and Fetal Alcohol Effects (FAE), which describe a distinct cluster of developmental anomalies resulting from alcohol-induced trauma to the fetus, are more common in many Indian communities than in the U.S. at large and reach unparalleled rates in some tribes, notably those of the southwestern Plains (May et al., 1983). The prevalences of FAS and FAE have been found to be 3 to 4 times higher in the youngest age groups (0–4 years of age), suggesting substantially increasing rates of occurrence. It is not surprising, then, that Fetal Alcohol Syndrome is the leading major birth defect among southwestern Indians (May, 1989).

There are no simple explanations for why Indian women drink, particularly in the face of these severe consequences. Reasons often given include cultural disruption, loss of social controls, prejudice, poverty, role reversals, peer group dynamics, and familiar socialization. Sudden and rapid culture change brought on by social, economic, and educational pressures has wrecked havoc with individual as well as collective identity among Indian people. Some scholars argue that the heavy drinking described above is in response to the consequently decreased sense of self-worth and alienation (Holmbred, Fitzgerald, & Carman, 1983). Change of this nature clearly alters traditional forms of social control, such that community sanctions, formal or informal, may no longer obtain (Medicine, 1982). It is also evident that drinking patterns introduced and reinforced through families and adolescent peer clusters become routinized in young adulthood (Medicine, 1982; Oetting & Beauvais, 1987). Hence, Indian women are readily incorporated into the ever-present drinking parties that can be found in reservations, adjacent towns, and inner city areas (Maynard, 1969; Weibel-Orlando, 1986; Weisner, Weibel-Orlando, & Long, 1984): social events once the sole province of Indian men. Lastly, Weibel-Orlando (1986) has pointed out that easy relief from parenting responsibilities—typically assumed by the extended family, and the fostering of a dependency mentality combine with this early introduction to chronic, heavy drinking to encourage a destructive cycle from which there is no easy escape.

The majority of Indian substance abuse treatment programs funded by the Indian Health Service (IHS), and typically delivered by tribal or urban Indian programs, have been established for or tend largely to attract men. Mail (1985) proposed that this may be due to several factors, including: (1) Indian alcoholism is perceived to be a male problem; (2) cultural expectations may prevent equal access to treatment (e.g., a woman should be home caring for her children); (3) tribal societies generally tend to be more protective of women, and (4) the majority of counselors are male, which may discourage some women from utilizing their services. Only four comprehensive programs funded by the IHS specifically target women: (1) The Ponca City Inpatient Treatment Program in Oklahoma; (2) The residential program for women of the Native American Rehabilitation Association in Portland, Oregon; (3) The outpatient support program of the Yakima Nation of Washington, and (4) The Wren House, a halfway house for women in Duluth, Minnesota. Most rural, reservation-based and urban programs seek to serve women clients, but only the few cited above are specifically designed to address gender-specific treatment issues and concerns. Examples include the provision of child care, attention to sexual victimization, and elimination of male-female role expectations.

Alcohol abuse/dependence prevention programs seem much more responsive in number and emphasis to the needs of Indian women. For example, a survey by Owan, Palmer, and Quintana (1987) of IHS Alcoholism/Substance Abuse Program Branch contractors revealed that 12% of the 999 preventive intervention activities carried out by them specifically targeted Indian women. Significant numbers of the remaining activities focused upon parents (12%), single parents (12%), pregnant women (8%), and older adults (7%), obviously encompassing large percentages of the female population. Program content stressed information, skills, and resources of immediate benefit to Indian women, in particular alcohol/substance abuse education (15%), building self-esteem and coping skills (14%), decision-making skills (13%), awareness of community resources (12%), developing self-help groups (11%), family bonding and enrichment (11%), and effective parenting (8%). Many of the contractors' concerns with respect to future program planning and development also included issues specific to Indian women, notably alcohol-related family violence and abuse (14%), fetal alcohol syndrome (13%), and alcohol abuse-related pregnancies (5%).

Interest in Indian women and alcohol has moved from the shadows of an almost exclusive concern for alcoholism among their male counterparts. The challenge, now, is to continue this inquiry on all fronts. We need to better understand not only the dynamics that place Indian women at special risk for alcohol abuse/dependence, but also to explicate the factors that protect many from it as well as assist in their recovery and continued sobriety. Answers to these questions lie at the interface of the social roles and obligations of being female in this unique culture.

REFERENCES

Holmgren, C., Fitzgerald, B. J., and Carman, R. S. (1983). Alienation and alcohol use by American Indian and Caucasian high school students. *The Journal of School Psychology, 120,* 139–140.

Mail, P. D. (1985). *A briefing book for the Alcoholism Program review.* Albuquerque, NM: Alcoholism Program Branch, Indian Health Service.

Manson, S. M., Shore, J. H., Baron, A. E., Ackerson, L., and Neligh, G. (in press). Alcohol abuse and dependence among American Indians. *In* J. E. Helzer & G. J. Canino, (Eds.), *Alcoholism—North America, Europe, and Asia: A coordinated analysis of population data from ten regions.* Oxford, England: Oxford University Press.

Manson, S. M., Shore, J. H., Bloom, J. D., Keepers, G., and Neligh, G. (1987). Alcohol abuse and major affective disorders: Advances in epidemiologic research among American Indians. *In* D. L. Spiegler, D. A. Tate, S. S. Aitken, and C. M. Christian, (Eds.), *Alcohol use among U.S. ethnic minorities.* NIAAA Research Monograph No. 18. Department of Health and Human Services Publication No. (ADM) 87–1435. Washington, DC: U.S. Government Printing Office.

May, P. A. (1989). Alcohol abuse and alcoholism among American Indians: An overview. *In* T. D. Watts and R. Wright, Jr., (Eds.), *Alcoholism in minority populations.* Springfield, IL: Charles C. Thomas Publishers.

May, P. A., Hymbaugh, K. J., Aase, J. M., & Samet, J. M. (1983). Epidemiology of fetal alcohol syndrome among American Indians of the southwest. *Social Biology, 30,* 347–387.

Maynard, E. (1969). Drinking as part of an adjustment syndrome among the Oglala Sioux. *Pine Ridge Research Bulletin, 9,* 33–51.

Medicine, B. (1969). The changing Dakota family and stresses therein. *Pine Ridge Research Bulletin, 9,* 1–20.

Medicine, B. (1982). New roads to coping—Siouan sobriety. *In* S.M. Manson, (Ed.), *New directions in prevention among American Indian and Alaska Native communities.* Portland, OR: Oregon Health Sciences University, pp. 189–212.

Moss, F. E. (1979). *Drinking attitudes and practices in twenty Indian communities.* Salt Lake City, UT: Western Region Alcoholism Training Center, University of Utah.

National Institute on Alcohol Abuse and Alcoholism. (1987). *Alcohol and health: Sixth special report to the U.S. Congress.* Washington, DC: U.S. Government Printing Office.

Oetting, E. R. and Beauvais, F. (1987). Epidemiology and correlates of alcohol use among Indian adolescents living on reservations. *In* D. L. Spiegler, D. A. Tate, S. S. Aitken, and C. M. Christian, (Eds.), *Alcohol use among U.S. ethnic minorities.* NIAAA Research Monograph No. 18. Department of Health and Human Services Publication No. (ADM) 87-1435. Washington, DC: U.S. Government Printing Office.

Owan, T. C., Palmer, I. C., and Quintana, M. (1987). *School/community-based alcoholism/substance abuse prevention survey.* Washington, DC: Alcoholism/Substance Abuse Program Branch, Indian Health Service.

Roy, C., Chaudhri, A., and Irvine, D. (1970). The prevalence of mental disorders among Saskatchewan Indians. *Journal of Cross-Cultural Psychology, 1*(4), 383–392.

Sampath, B. M. (1974). Prevalence of psychiatric disorders in a southern Baffin island Eskimo settlement. *Canadian Psychiatric Association Journal, 19,* 303–367.

Shore, J. H., Kinzie, J. D., Hampson, D., and Pattison, M. (1973). Psychiatric epidemiology of an Indian village. *Psychiatry, 36,* 70–81.

U.S. Bureau of Census: *A Statistical Profile of the American Indian Population: 1980 Census.* Washington, D.C.: U.S. Government Printing Office, 1984.

U.S. Department of Health and Human Services. (1988). *Indian Health Service Chart Series Book.* Washington, DC: U.S. Government Printing Office.

Weibel-Orlando, J. (1986). Women and alcohol: Special populations and crosscultural variation. *In Women and alcohol: Health-related issues.* NIAAA Research Monograph No. 16. Department of Health and Human Services Publication No. (ADM) 867-1139. Washington, DC: U.S. Government Printing Office.

Weibel-Orlando, J., Long, J., & Weisner, T. (1984). Urban and rural Indian drinking patterns: Implications for intervention policy development. *Substance and Alcohol Actions/Misuse, 5,* 45–57.

Weisner, T. S., Weibel-Orlando, J., and Long, J. (1984). "Serious drinking," "White man's drinking," and "teetotaling": Drinking levels and styles in an urban American Indian population. *Journal of Alcohol Studies, 45,* 237–250.

Whittaker, J. O. (1962). Alcohol and the Standing Rock Sioux tribe. I: The pattern of drinking. *Quarterly Journal of Studies on Alcohol, 23,* 468–479.

Whittaker, J. O. (1982). Alcohol and the Standing Rock Sioux tribe: A twenty-year follow-up study. *Journal of Studies on Alcohol, 43,*(3), 191–200.

CHAPTER 17

Lesbian Women

Dana G. Finnegan, Ph.D., CAC
Emily B. McNally, Ph.D., CAC

A review of the alcohol and drug literature reveals a serious lack of articles and books about women and lesbians, as well as flaws in what literature does exist (McNally, 1989). Several books about women alcoholics include minimal discussion about or mention of lesbians (Peluso and Peluso, 1988; Sandmaier, 1980; Wilsnack and Beckman, 1984). These books also contribute to some major perceptual problems about lesbian alcoholics. One of these is that of viewing lesbians as belonging to a category which is separate and distinct from other women. Yet there are lesbians in every category of women alcoholics: professional women, housewives, grandmothers, teenage girls, and homeless women; Black, white, Hispanic, Asian; rich, poor, educated, uneducated, married, unmarried, old, and young women. Most of the articles, chapters and books about these categories of women ignore the issues and concerns of lesbians.

Ignoring lesbians in these other categories leads to another perceptual problem—that there is "a lesbian alcoholic" who belongs to a separate and distinct category, different from other women. This perception tends to perpetuate the stereotype of "the lesbian" as a young, poor, white, aggressive, masculine, man-hating woman wearing combat boots and a leather jacket. Related to this problem of stereotyping is that of seeing their problems as caused by their being lesbians. They are not seen as having problems with jobs, relationships, low self esteem, abuse, denial, shame, feelings, and other issues common to most recovering alcoholic women.

The literature on alcoholic women is also flawed by its failure to directly address issues of sexual identity, sexism, and homophobia. A number of writers discuss such topics as women alcoholics' shame issues, sex role conflicts, and affiliation needs without ever considering the powerful influences of society's homophobia and sexism on these women (Beckman, 1978; Gomberg, 1987; Moyar, 1988). In this context of homophobia and sexism, women are taught to be fearful and ambivalent about affectional and sexual feelings toward other women. Many women experience shame, confusion, and fear about their sexual/affectional identity during their drinking (McNally, 1989). Unfortunately, most of the alcoholism literature ignores both the oppressive context and women's struggles with their feelings. For example, Moyar (1988) talks about how important it is to their recovery that alcoholic women's affiliation needs be met. She contends that women's relationships with other women are the vehicle for meeting these needs, but she never mentions what effects homophobia might have on these relationships.

Perceptual problems reach beyond the alcohol and drug literature. Prevention and treatment professionals often do not ask women about their sexual and affectional feelings, experiences, and identity. Instead, too many of them are hampered by narrow perceptions of who "the lesbian" is. If she doesn't fit the stereotype, she must not be a lesbian. Thus, many lesbians are not identified and are not given the care they need. They are at best ignored and at worst "erased." They are made invisible, and are not reached, and are not helped.

Alcoholism professionals need to provide constructive help to every alcoholic woman who must deal in her recovery with sexual identity issues. To do so, professionals must create a safe and nurturing atmosphere in which women can explore their feelings and experiences. Otherwise, the problems will continue to exist.

The problems are twofold. Women who fit the stereotype or women who identify themselves as lesbians risk, and oftentimes pay, a great price. They risk being misunderstood or patronized or feared; they risk rejection, contempt, ostracism. Or they may be tolerated or pitied. With negative experiences and without strong and positive support, many women may not recover.

On the other hand, the greater majority of lesbian alcoholics will not willingly risk the price of self disclosure and will remain invisible. In order to survive in a homophobic atmosphere, they will "pass" as heterosexual. But this survival mechanism tends to cut them off from help and support for their real selves. They may feel isolated, alienated, unique, shamed, and lonely. If they survive treatment, they are likely to be at risk for relapse. Although they have received help with their alcoholic identity, they are left alone with all the issues and problems related to dealing with a lesbian identity in a homophobic and sexist world.

Unless professionals learn to look for and to see hidden, invisible lesbian alcoholics; and unless they learn how to reach out to lesbian alcoholics, support them, and help them feel safe, the professionals cannot adequately help these women to recover. Instead of offering solutions, the alcoholism professionals may become part of the problem.

THE SEVERITY OF THE PROBLEMS

Incidence

The incidence of gay men and lesbians who are at risk for alcoholism or are alcoholic is usually cited as between 28 and 32 per cent (Fifield, 1975; Lewis, Saghir, and Robins, 1982; Lohrenz, Connelly, Coyne, and Spare, 1978). Although McKirnan and Peterson's study (1987) did not establish support for such high figures, they did find that "rates of alcohol problems are higher than the general population among . . . in particular, homosexual women, indicating some additional risk for substance abuse" (p. 21).

Another of their findings is that "the homosexual sample showed a high rate of alcohol use in different recreational settings and in response to stress, consistent with anecdotal reports about this population" (p. 21). The major public recreational setting for lesbians has traditionally been the lesbian bar. Although often sleezy and grimy and although often situated in unsafe neighborhoods, these bars offer islands of relative safety in which lesbians can be themselves and meet other lesbians (Abbott and Love, 1972; Jay, 1978).

The participants in McNally's (1989) study of lesbian recovering alcoholics indicated that some of them drank to *be* lesbian and some of them drank to *not* be lesbian. Most of the women drank in order to cope with their internalized homophobia, with their fear, and with their shame—or as McKirnan and Peterson (1987) describe it, in response to stress. Thus it would seem that lesbians who drink to cope with being lesbians are at risk.

Triple Stigmas

Much of the current literature points out that women alcoholics have special issues and problems just because they are women. According to Schur (1984), being female in this society is a deviant status. Women are inferior outsiders. Coupled with this inferior status are all the feelings and conditions created by it, such as low self-esteem, conflicts about stereotyped sex-roles, isolation, shame, lack of social status and economic resources, abuse, lack of job skills. If a woman becomes alcoholic, she faces the culture's stigmatizing attitudes toward women "drunks." They are seen as shameful—promiscuous, "bad" women (Beckman, 1978; Blume, 1986, 1989; Gomberg, 1987, 1989; Wilsnack, 1984). These can be huge stumbling blocks to recovery and most alcoholic women have a very difficult time because of one or more of these issues.

Lesbian alcoholics' experiences with these stumbling blocks created by the double stigma of being a woman and being an alcoholic are compounded by the third stigma of being lesbian. In a homophobic society such as this one, to be lesbian is to be deviant, an outcast (Anderson & Henderson, 1985; Finnegan & McNally, 1987; McNally, 1989; Underhill, 1981; Vourakis, 1983). Thus to be a woman, an alcoholic, and a lesbian is to be triply stigmatized, triply traumatized, conditions which can make recovery extremely difficult.

Recently a lesbian, an attractive young model who does not fit the stereotype, described what alcoholism treatment was like for her in an expensive, 28-day in-patient program. She had decided not to "come out" unless she knew she was in a safe place. Then she began to hear homophobic remarks about another patient who was "out" as a lesbian, remarks which greatly distressed her. She decided to "come out" in order to support the other lesbian. When the other patient left treatment early, this young woman felt so alone and afraid that she asked the staff for help. Their response was, "You'd better get used to it. That's the real world." She had heard these same words as a young girl when she told her mother about being sexually abused by her stepfather. Since she did not have alcohol to cope with the pain and isolation she felt, she found another way to cope with these violations of her self esteem. She became bulimic. Although she has remained sober since that treatment experience, she is now struggling with a serious eating disorder. Is this what is called "successful" alcohol treatment?

Whether lesbians hide their sexuality during treatment or "come out" they may end up feeling, and being, isolated and alone. If "successful" recovery is measured by the stopping of drinking, then they may even be rehabilitative successes. But usually, they are left with many of the stumbling blocks that most alcoholic women have. And they are left alone with them.

Unfortunately, horror stories about lesbians in alcoholism treatment such as this one are not rare. Some women have been told that their lesbianism is caused by their alcoholism or vice versa; some have been told that their lesbianism is a phase they will outgrow when they get sober; some have been refused treatment or discharged from treatment; some have been warned not to "attack" other females; and others have met with revulsion, curiosity (or voyeurism), or hostility. It is a tribute to many women that they have managed to recover despite such negative experiences.

WHAT NEEDS TO BE DONE?

The alcoholism field must address a number of issues in order to help lesbians or women with sexual identity concerns recover from their addictions. Many of the necessary changes and actions center on making treatment safe enough for these women to recover. "Safety" may be difficult to define because it varies from woman to woman and time to time. For some women it may mean talking about their lesbian identity and being accepted by others as lesbians, while for others it may mean putting sexual identity aside for a period of time while they focus on sobriety (McNally, 1989). Counselors need to be able to identify where a woman is in her sexual identity development in order to provide her with safety. In addition, the book, *Dual Identities* (Finnegan & McNally, 1987) presents an "institutional audit" which can be used to check for homophobic practices in alcoholism treatment centers for those interested in making their programs safer and more sensitive

Although there seems to be some variability in what feels safe to different women at different times, some women have talked about practices in many treatment programs that do *not* feel safe. Being one of several women in a treatment program with a majority of men does not feel safe to most women. Being a lesbian who is made to disclose her lesbian identity in a group of heterosexual men and women does not feel safe to most lesbians. Having to talk about sexual experiences and sexual abuse in a mixed group does not feel safe to most women. Hearing homophobic remarks made by other patients and staff does not feel safe. Sometimes homophobia is particularly painful and shaming when it takes the form of laughter and joking about closeness between two men or two women. Hearing the word "straight" used to describe being drug-free does not feel safe to most lesbians. Many lesbians describe enduring un-safe and sometimes abusive alcoholism and drug treatment because they were so desperate to get sober that they feared leaving, being out on the street, and dying if they did not complete treatment.

PREVENTION

Despite the high incidence of alcohol and drug abuse in the lesbian community, prevention programs specifically for lesbians are almost nonexistent. Prevention programs are needed particularly to address alcohol problems in high risk groups, such as adolescent girls and women in their early twenties. Sex education programs in the schools should not assume that all children growing up are and will be heterosexual. They should provide factual information about the variety of ways human beings express their sexual and affectional feelings and identities. It is during adolescence that many girls begin struggling with sexual identity issues or are coming out for the first time as lesbians. Many of them are frightened and do not know how to be sexual, social, or lesbian without alcohol (McNally, 1989).

Many recovering alcoholic lesbians say they began experiencing problems with alcohol during early adolescence. They say that alcohol became linked with acting out behaviors, getting into trouble, making suicide attempts, and feeling frightened and confused about their sexual and affectional feelings. Prevention programs should target girls who have just begun to get into trouble, before these girls have to experience years of problems and progression of their alcoholism. Professionals should help the girls look at their drinking and explore their sexual identity. They should give them acceptance and support, as well as information and choices.

Many lesbians still associate "coming' out with bars and drinking. It would be helpful if there were other alternatives for girls and women who are exploring their feelings and identities. If being "chemically free" and not drinking were acceptable ways to attend dances, concerts, and other activities, women might be able to associate being sober with having fun and getting to know themselves and others.

Some lesbians recovering alcoholics say that they had no one to talk with about their feelings and confusion about possibly being lesbian. They turned to alcohol to cope, in part, because they had no where else to turn. It would be helpful if the "gatekeepers," the people who come in contact with girls and women who might be struggling with sexual identity concerns, were trained to identify early stages of alcoholism and the signs indicating confusion about sexual identity development. These gatekeepers are people such as resident assistants in college dormitories, guidance counselors in junior and senior high schools and colleges, coaches of girls' sports, camp counselors, staff in women's centers, and all types of health and mental health professionals.

TREATMENT

One of the most important issues in alcohol and drug treatment programs is that of safety, training, and sensitivity. Making treatment safe for lesbians and women who are afraid they might be lesbians must be a goal. It is difficult to imagine how any treatment program can do this when the staff do not feel safe enough to be open about themselves. In many treatment centers, nurses, doctors, counselors, social workers, and administrators are afraid to let anyone know about their sexual identity for fear of being rejected, fired, or treated negatively by others. The message given to patients is that staff is not safe; therefore patients are not safe.

Even in treatment centers where staff are safe enough to be open about themselves, all staff need to be adequately trained in sexual identity issues and helped to work through their own homophobia and sexism. They should be given on-going training, supervision, and consultation in working with their lesbian patients. They should understand the stages of sexual identity transformation in order to assist their patients. For example, they should be able to know how to ask questions in a sensitive manner to gather the information they need to help their patients. A woman coming into treatment may have been sexual with other women while she was drinking, but may not call herself a lesbian and may be very frightened to be seen as a lesbian. Counselors should learn to respect their patients' ambivalence, shame, and fear.

Asking good, sensitive questions is part of providing safe and supportive treatment to patients. Another important part of quality treatment is responding to patients' questions, feelings, and confusion. Staffs should know how to answer factual questions and how to refer the patient to other resources, such as books, groups, and professionals.

For the past decade these issues have been presented, though there has been some positive progress in making changes, much has unfortunately remained the same. There is a need for radical improvements in prevention and treatment for lesbian alcoholics and for all women alcoholics who must deal with sexual identity issues. If women are to receive the quality treatment they deserve, then the alcoholic field must make and keep a commitment to create safety and support for staff and patients alike. Professionals must commit to getting training, examining and changing their attitudes, exploring their own sexual identity issues, and providing non-sexist, non-homophobic help to *all* women.

REFERENCES

Abbott, S., and Love, B. (1972). *Sappho was a right-on woman: A liberated view of lesbianism.* New York: Stein and Day.

Anderson, S. C., and Henderson, D. C. (1985). Working with lesbian alcoholics. *Social Work, 30*(6), 518–525.

Beckman, L. J. (1978). Self-esteem of women alcoholics. *Journal of Studies on Alcohol, 39*, 491–498.

Blume, S. (1986). Women and alcohol: A review. *Journal of the American Medical Association, 256*(11), 1467–1470.

_____. (1989, January/February). Why we are concerned: An overview of women and alcohol. *The Counselor,* pp. 14, 20.

Fifield, L. (1975). *On my way to nowhere: Alienated, isolated, drunk.* Los Angeles: Alcoholism Center for Women.

Finnegan, D. G. and McNally, E. B. (1987). *Dual Identities: counseling chemically dependent gay men and lesbians.* Center City, MN: Hazelden.

Gomberg, E. L. (1987). Shame and guilt issues among women alcoholics. *Alcoholism Treatment Quarterly, 4*(2), 139–155.

_____. (1989, January/February). Shame and guilt. *The Counselor,* pp. 23–24.

Jay, K. (1978). A journey to the end of meetings. In K. Jay and A. Young (Eds.), *Lavender culture* (pp. 452–457). New York: Harcourt Brace Jovanovich.

Lewis, C. E., Saghir, M. T., and Robins, E. (1982). Drinking patterns in homosexual and heterosexual women. *Journal of Clinical Psychiatry, 43,* 277–279.

Lohrenz, L., Connelly, J., Coyne, L., and Spare, K. (1978). Alcohol problems in several Midwestern homosexual communities. *Journal of Studies on Alcohol, 39,* 1959–1963.

McKirnan, D. J. and Peterson, P. L. (1987, September). *Alcohol and drug use among homosexual men and women: Descriptive data and implications for AIDS risk.* Paper presented at the American Psychological Association conference, New York.

McNally, E. B. (1989). *Lesbian recovering alcoholics in Alcoholics Anonymous: A qualitative study of identity transformation.* Unpublished doctoral dissertation, New York University, New York.

Moyar, M. (1987). Female alcoholism and affiliation needs. In M. Braude (Ed.), *Women, power, and therapy* (pp. 313–321). New York: Harrington Park Press.

Peluso, E., and Peluso, L. S. (1988). *Women & drugs: Getting hooked, getting clean.* Minneapolis: CompCare.

Sandmaier, M. (1980). *The invisible alcoholics: Women and alcohol abuse.* New York: McGraw Hill.

Schur, E. M. (1984). *Labeling women deviant: Gender, stigma, and social control.* Philadelphia: Temple University Press.

Underhill, B. L. (1981, July). *Elements of effective services for the lesbian.* Paper presented at the Women in Crisis Conference, New York.

Vourakis, C. (1983). Homosexuals in substance abuse treatment. In G. Bennett, C. Vourakis, and D. S. Woolf (Eds.), *Substance perspectives* (pp. 400–419). New York: Wiley.

Wilsnack, S. (1984). Drinking, sexuality, and sexual dysfunction in women. In S. Wilsnack and L. Beckman (Eds.), *Alcohol problems in women* (pp. 189–227). New York: Guilford Press.

Wilsnack, S. and Beckman, L. (Eds.). (1984). *Alcohol problems in women: Antecedents, consequences, and intervention.* New York: Guilford Press.

CHAPTER 18

Co-Dependency and Dysfunctional Family Systems
Sharon Wegscheider-Cruse, M.A.

Co-dependency is a relatively new term used to describe the condition affecting persons who have lived in dysfunctional situations. It has varying definitions. My own is:

> Co-dependency is a toxic relationship to a substance, a person, or a behavior that leads to self-delusion, emotional repression, and compulsive behavior that results in increased shame, low self-worth, relationship problems, and medical complications.[1]

Co-dependency is a disease. It is a specific condition characterized by preoccupation and extreme dependence on another person (emotionally, socially, sometimes physically), or on a substance (such as alcohol, drugs, nicotine, and sugar), or on a behavior (such as workaholism, gambling, compulsive sexual acting out). This dependence, nurtured over a long period of time, becomes a pathological condition that affects the co-dependent in all relationships.

Co-dependency, like alcoholism and drug abuse, is not owned by any particular race, class, or cultural group. Anyone can be affected. However, we do know that certain individuals are more likely than others to be affected. Included in this high-risk group are adult children from alcoholic or other dysfunctional families.

In alcoholic families particularly, the dysfunctional system requires that family members fulfill certain roles.[2] The person assuming each role may change from time to time, and one person may take on more than one role, but the functions of the roles always need to be carried out to support the system.

These roles are:

- *The Dependent*, whose addiction is the family preoccupation.

---

1. Wegscheider-Cruse, S. (1988). *The Co-Dependency Trap.* Nurturing Networks, Inc.: Rapid City, SD.

2. Wegscheider-Cruse, S. (1981). *Another Chance.* Science and Behavior Books, Inc.: Palo Alto, CA.

157

- *The Enabler,* who supports the Dependent by assuming responsibilities and protecting them from consequences of their behavior. This is most often the spouse.
- *The Family Hero,* often the oldest child, whose overachievement and responsibility provide self-worth for the family.
- *The Scapegoat,* whose defiance and acting out help take the focus away from the underlying family dysfunction.
- *The Lost Child,* lonely and quiet, who provides relief by being the child no one has to worry about.
- *The Mascot,* whose clowning and humor provide comic relief and distraction from the family pain.

All the children in such a dysfunctional family system learn to hide and deny their feelings. They learn to mistrust their own perceptions of reality because what they feel, hear, and see fails to match what they are told. All of them learn, "I am less important than Dad's or Mom's drinking. I don't matter." By the time they are adults, they almost certainly will have developed the symptoms and complications of co-dependency.

SYMPTOMS

Symptom One: Delusion (Distorted Thinking)

Delusion is when we allow our thoughts to get distorted by not taking in all the information that is there. Instead, we live with our own narrow view of what we see and how we see it.

In unhealthy families, people are taught either directly or indirectly not to be honest with all they see, hear, feel. And what they begin to learn in a painful family is how to separate themselves from that total view and live with a limited view. Disassociating from the reality of what we truly feel, is being dishonest with ourselves. So, distorted thinking is lying to yourself and to others about the pain you're in. It is not experiencing the whole picture.

This minimization and rationalization of the co-dependent is often deeply felt and truly believed—a "sincere delusion." Dependency is a perceived experience—one which grows out of an individual's compulsive response to something that is so safe, so self-worth producing, so comforting that he does not want to (can't!) be without it. Dependence, whether on alcohol, drugs, money, food, or another person, is dependence all the same. And it is maintained through denial. Confusion about making decisions and rigid, judgmental attitudes (when everything is black or white, with no shades of gray) are expressions of the denial/delusion aspect of co-dependency.

Symptom Two: Emotional Repression (Distorted Feeling)

Each time we disassociate from reality and deny what is happening, we repress or stuff the feelings that go with the event. Shutting down those feelings becomes a condition of emotional repression. After while, as time goes by, those feelings that one shuts down and doesn't feel become all mixed together. This emotional mixture is called undifferentiated emotion. That means we don't even know what we feel anymore. Is it anger, guilt, hurt, shame? What is it? The events get lost. First of all, we are denying the events. And, over time, the events get lost and people experience free-floating feelings (free-floating anxiety).

The emotions buried deep inside—fear, guilt, shame, anger, loneliness, hurt—are potentially immobilizing. People can work through these feelings and eventually get past them as they recover health and hope—or they can repress these feelings. Life becomes a social theater, and people wear masks. Some examples of the feelings hidden behind those masks are:

1. Anger. Often, the co-dependent's attempts to control backfire—attempting to control someone else is a self-defeating project. This results in frustration, which turns to anger as things fall apart. The anger turns to rage, which is often swallowed or stuffed. Store enough feelings, and the symptom that appears is depression. As time goes on, the mind becomes a storehouse for pent-up memories and hidden resentments. One may think these hurts have disappeared, but they have not. Not after weeks, months, or even years.

    This hidden anger and resentment lead to a suppressed experience of love and joy throughout one's life. When one is fighting back painful feelings, current relationships are affected in a variety of ways. It is difficult to show affection for loved ones if it also feels necessary to maintain control by keeping people at arm's length. The sad result of avoiding anger is just the opposite of what people expect. The individual is capable only of in authentic forms of human interaction. Eventually relationships become difficult to handle, reach an impasse, and fall apart.

2. Loneliness. The co-dependent feels as though nobody really understand. Feelings of uniqueness are nurtured by growing isolation. This fosters further loneliness and isolation.

3. Fear. Co-dependents live with feelings of low self-worth, believing themselves to be totally powerless. Yet their efforts to get a handle on as many situations as possible result in the co-dependent becoming a controller. The pathological need to control so much naturally results in a great deal of fear. There is fear of being found out, of not being good enough, of relaxing and letting one's guard

down because things might fall apart. There is the all-consuming fear of abandonment, loneliness and rejection.
4. Inadequacy. Co-dependents tend to compare external appearances. There is little in the way of true sharing, intimacy, and emotional expression. Most relationships are limited to superficial talk about the weather, jobs, etc. Even the sharing of seemingly intimate sides can be just performance. Co-dependents are skilled performers. They live their lives like a dress rehearsal for a play that has not yet opened. Going to school, marrying the right person, having a baby, etc., are merely preparatory for the co-dependent. This keeps the individual from feeling good about the achievement of smaller goals as ends in themselves. It also may prevent the accomplishment of greater goals.
5. Hurt. After the individual has trapped himself in the caretaker role, it becomes expected. Family and friends take that responsibility for granted, and this can be painful for a person who has expected at least the reward of abject gratitude for their super-responsibility. Living to take care of others is a joyless existence; and acknowledging the trap can hurt so much that the co-dependent will resist information and feedback in order to maintain the illusion that they are needed as much as they wish they were. Under many a super-responsible person is a sad, lonely, hurting one.

Symptom Three: Compulsive Behavior (Distorted Behavior)

Over the years, for children from painful homes, there has been an inner cesspool or emotional abscess of shame, hurt, loneliness, anger, inadequacy, sadness and hopelessness. It is deep inside and kept locked away. It hurts, and the need for relief from the pain becomes a craving.

The first stage of dependency is to seek some sort of change of mood that is pleasurable or relief-producing. Different substances and behaviors work for different people. Some find one of the predictable mood-changing substances such as alcohol. If the person is set up genetically to react to mood-changing chemicals, dependency and addiction begin.

The chemicals provide relief from that deeply-buried inner pain. It is a temporary relief, but for those whose bodies react addictively, it is a potent one which works every time. Street drugs, prescription drugs, and alcohol are all effective pain-relievers. So is nicotine.

Smoking interferes with the attainment of intimacy and personal growth. It serves as an insulator from the world of uncertainty and psychic pain.

Now, not everybody is set up to be able to get relief from alcohol and drugs. It just doesn't work for them because they do not have the genetic make-up for it. So there are many more people who have lived in painful homes who need relief and who have this carving than can successfully use drugs and alcohol.

These people find other things, depending on their circumstances, depending on their family system, depending on what works for them. For families that are very perfectionistic, stoic and cognitive, people are set up to get a rush through the power of control. In those kinds of families, we may find the workaholic rush. These people stay at a certain tolerable level of their internal pain by staying frenetically active. This family system may also produce the anorexic, the person who starves and gets a rush from being able to turn away food. Some people are set up to be sugar-sensitive, and for them, sugar does what alcohol does not do. Eating disorders are a chronic, progressive and sometimes fatal disease if left untreated.

People can get their rush or relief with what we call "green-paper" addiction. That moves all the way from occasional overspending to living on credit cards, never really believing you have to pay them off. Green-paper addiction includes gambling, which is one of the toughest of the compulsions to treat.

So drugs, alcohol and nicotine are major chemical invaders, the most common chemical substances that we see people form toxic relationships with. Then some of the other toxicities are in behaviors . . . the food disorders, workaholism, and relationship dependency.

Complication One: Low Self-Worth (Shame)

The first complication of co-dependency is chronic low self-worth (a condition of feeling "shame"). The difference between guilt and shame is:

1. Guilt is felt when one does something to harm oneself or others. "I feel badly about something." The wonderful thing about guilt and the reason guilt can be such a positive feeling is that when I recognize it, I can make amends and then feel good about myself. Gilt has a pattern that brings us back to our good feeling if we pursue it. Every piece of guilt can be amended one way or another. So guilt is a very productive feeling. To begin to feel guilty in places where it's appropriate and then begin to make amends is to be on a straight path to self-worth and healing.

The difference between guilt and shame is that guilt is the feeling of "I've done something bad and I would like to make amends about it and change."

2. Shame is feeling that I am what is bad. I am bad, not that I did something bad, I just am faulty. I am bad. Dysfunctional families tend to produce shame-based people. In hurting families, people bring genera-

tions of shame into their current life. What does not get resolved in the generation before comes as a package. It's part of the shame.

Then, when you add all those old family rules, those inhuman rules, what you end up with is people who feel "less than" . . . "unworthy." They end up with what I call chronic low self-worth, which is being shame-based. It's the same thing. When we are shame-based, then it is very difficult to make decisions in behalf of what we need for ourselves. People get into this place of perceived powerlessness. They believe that they are bad and unworthy and there is nothing they can do to change.

Complication Two: Relationship Problems

The co-dependent lives with environmental wars. The smallest unit is the couple. It is rare to find relationship satisfaction when one is emotionally frozen and behaviorally compulsive. Knowing and responding to a partner is simply not the major focus. Beyond the coupling, the family, friends, the job and work environment all suffer from lack of focus and commitment.

Learning to clearly face the reality of hurting relationships in the first step of recovery. Next comes the behavioral rebuilding of relationships one values. Some toxic relationships need to end, and others need to be rebuilt. All relationships deserve attention in co-dependency recovery.

Complication Three: Medical Complications

There's one more complication. If, through low self-worth and shame we do not make changes, we will stay stuck. Stuckness is another word for a subtle death wish. People who are stuck are beginning the dying process. But early on, what happens in this stuckness is that we start having medical problems. When we will not reach out and ask for what we need and get ourselves nourished in ways that we need to be fed, then our bodies go ut of harmony, out of alignment, out of ease and they become diseased. We make ourselves sick.

When we are functioning in harmony and perceive ourselves as high functioning, we'll stay healthier. When we perceive ourselves as low energy, stuck, down, powerless, we become more susceptible to illness.

The good news is that co-dependency is treatable.

Effective treatment includes:

1. Confronting self-delusion with new information; with learning comes understanding.
2. Creating a safe atmosphere where feelings can surface to be shared and discharged so healing can take place.
3. Providing atmosphere where it is safe and possible to detach from compulsive behavior.

The co-dependent does best with a two-part recovery program. One part is involvement in a 12-step group, and the other part is a relationship with a co-dependency therapist or a professionally-led group.

I recommend that people pick one-or two-step programs to attend that fit with the major compulsion that they are struggling with. They then should also reserve time and energy to invest in a therapeutic process that is intense and behavior-change oriented. When one picks a therapist to work with in co-dependency recovery, it's important to choose someone who thoroughly understands the disease concept of co-dependency. If that person has come from a painful family of his/her own, it is crucial that he/she has already received and healed in his/her own treatment process.

Therapy can come in many forms. It can be an outpatient program. It can be a short, intense program, 8 to 10 days. What is important is that the program be specifically designed and facilitated by people who are specifically trained in co-dependency work. Therapy can and does come to an end.

Continued involvement in a 12-step program can be a much longer commitment. The analogy I sometimes use is that a good treatment program is like emotional surgery. It is a place to do a lot of work all at once. Emotional surgery is then followed by long-term healing, which takes place in 12-step support groups.

This approach can lead to full recovery, getting to a place of "being" instead of always "doing". One can become fully alive, aware, and invested in day-to-day reality, meaning, and joy.

CHAPTER 19

Domestic Violence
N. Ann Lowrance, M.S.

Historically, family violence and chemical dependency have been linked. While commonalities exist, available data does not support the contention that substance abuse causes men or women to batter or that family violence causes substance abuse. When substance use disorders and domestic violence co-exist in the same family unit, both issues must be addressed.

AN OVERVIEW

The F.B.I. has identified domestic violence as the most frequently occurring and most underreported crime in the nation. Although the exact incidence is unknown, a recent national study of family violence found that severe physical violence is a chronic feature in at least 16% of American marriages (Browne, 1987). Every 15 seconds a woman is physically assaulted in her home (Dutton & Painter, 1981).

In the field of family violence, a battered woman is most commonly defined as any women who is repeatedly subjected to any physical or psychological behavior by a man in order to coerce her into doing what he wants without regard for her rights as an individual (Eberle, 1982). The dynamics of violent relationships are complex and involve a number of psychological and environmental factors.

Battering incidents generally involve a combination of physical and psychological assaults. Walker emphasized that most of the women she studied described incidents involving psychological humiliation and verbal harassment as their worst battering experiences, whether or not they had been physically abused (Eberle, 1982). Even in cases where abuse had consisted only of verbal or psychological behaviors, Walker found that the threat of physical violence was always present: each [woman] believed the batterer was capable of killing her or himself.

BATTERED WOMAN SYNDROME

Walker defined a three-phase cyclical pattern in battering relationship that is generally repeated. Battered women exhibit unique clusters of behavioral patterns in response to this cycle (Walker, 1979).

The initial phase of the cycle of violence, called the "tension reduction" phase, is characterized by arguments and minor battering incidents. The battered woman attempts to calm the batterer and often accepts unjustified blame to avoid his anger. This phase is characterized by rationalization minimization and denial. When tensions begin to escalate at the end of the tension reduction phase, the battered woman is most likely to seek help (Walker, 1979).

The second phase is the "acute battering" phase and is characterized by a lack of control and major destructiveness. The woman is subjected to severe physical and verbal abuse and is often seriously injured. Following the battering both the battered woman and the batterer experience shock and disbelief and tend to rationalize and minimize the seriousness of the attack. In recalling the battering incident, both report time distortion, dissociation, memory of certain details, a sense of disbelief, and a feeling of relative clam. It is because of this sense of shock that battered women usually do not seek help for 24 to 48 hours following the battering incident (Walker, 1979).

The third phase is known as the "loving contribution" phase. During this phase the batterer becomes remorseful, apologetic, and loving, and assures the woman that the battering will not be repeated. It is at this time the batterer is most likely to get help, but he often loses his resolve, especially if she has left him and then returns. His contrite behavior reinforces the woman's commitment to the relationship and her hope for change. Because his behavior during this phase is very loving, there is a tremendous drop in tension, and both diminish the severity and danger of the beating. Initially, the "loving contrition" phase magnifies the battered woman's hope and sustains the relationship. This phase tends to decrease in both length and intensity over time, which may eventually strengthen the woman's resolve to exit the violent relationship (Walker, 1979).

The cycle tends to be repeated. Episodes of batterers' violent acts increase in frequency and severity over time, which is different than other violent offenders who become less violent with age (Walker, 1979).

Stereotype Images of Family and Male-Female Relations

Battered women tend to have rigidly stereotypic images of marriage, family, and the roles of men and women within the family (Fojtik, 1977–78). They tend to believe that family life should center around the man, and that the man has the right to control them and their children.

He is responsible for making all major family decisions, while she is responsible for making the family "happy and healthy". Because of these ideals about marriage and the role of the wife, the battered woman feels as if she is a total failure. Her poor self-image is enhanced by the batterer's consistently blaming her for any family problem. These unrealistic expectations about families frequently lead battered women to overestimate the benefits of keeping the family intact. Many battered women report that they stayed with their abusers in order for their children to have two parents (Fojtik, 1977–78).

Traumatic Bonding

Traumatic bonding refers to "the strong emotional ties [which develop] between two persons where one person intermittently harasses, beats, threatens, abuses or intimidates the other." These ties which "manifest themselves in positive feelings and attitudes by the subjugated party for the intermittently maltreating or abusive party" have been observed between hostages and their captors, battered children and their abusive parent, and prisoners and their guards (Friedman, 1988, summer).

There are two features of social structure which are common in these relationships. One common feature is an imbalance of power. In the battering relationship, the battered woman perceives herself as less powerful than the batterer. This perception leads to dependency on the batterer, and the battered woman becomes "more negative in [her] self-appraisal, more incapable of fending for [herself], and thus more in need of the high power person" (Friedman, 1988, summer). The second common feature is the periodic nature of the abuse. In battering relationships, the woman is subjected to intermittent periods of abuse which are alternated with periods during which her husband treats her lovingly. "Such intermittent maltreatment has been found to produce strong emotional bonding effects in both animals and humans" (Friedman, 1988, summer).

Learned Helplessness

A common psychological result of the battered woman's experience is the development of "learned helplessness". "Battered women are repeatedly and unpredictably subjected to severe physical and psychological abuse over which they have no control and from which they generally cannot escape. Moreover, they usually find that there is little, if anything, they can do to alter the relationship or prevent further abuse" (Ewing, 1987). Battered women have generally tried to change the violent behavior of her abuser or escape the violent relationship by accessing all of the alternatives known to her. She has tried to alter her behavior in ways that the batterer has suggested. She may have talked with friends or family members. She may have sought advice from the family minister, priest, or rabbi. Generally, as the violent behavior escalates, she may seek protection from the criminal justice system. Because society in general lacks knowledge about domestic violence, she may receive inappropriate or ineffective response from these sources. As a result, the battered woman "believes or has learned that [s]he cannot control those elements of [her] life that relieve suffering, bring gratification, or provide nurture—in short, [s]he believes that [s]he is helpless. Eventually, because she believes there is no escape, the battered woman resigns herself to live in the violent relationship" (Labell, 1979).

Fear

Fear is the most common emotion experienced by the battered woman. She knows that the batterer is capable of violence because of her history with him. She may have tried to escape before and have been severely beaten afterwards. As a part of the psychological battering, abusers frequently threaten to harm her or her children if she tries to leave. So she is afraid for her own safety and the safety of her children.

This fear is justified. One-fourth of all the nation's homicides were committed by family members of the victim (National Coalition Against Domestic Violence National Toll-Free Hotline 1-800-333-7233). Nearly half of these intrafamilial homicides were between partners. "Of these, the majority of victims were women: Two-thirds were wives killed by their husbands, and one-third were husbands killed by wives. Women don't usually kill other people; they perpetrate less than 15 percent of the homicides in the United States. When women do kill, it is often in their own defense. A report by a government commission on violence estimated that homicides committed by women were seven times as likely to be self-defense as homicides committed by men" (Browne, 1987). Research over the last ten years indicates that women who leave their batterers are at a 75% greater risk of being killed by the batterer than are those who stay (Dutton & Painter, 1981). This leads to the conclusion that even though the battered woman may leave the abuser, her safety is not insured.

Suicide rates among both the abuser and the battered woman are high. One-third of all suicide attempts by women are a direct result of domestic violence (Browne, 1987). These attempts are due to the battered woman's sense of helplessness and hopelessness. Sixty-one percent of the men in Browne's study had threatened to commit suicide, generally in connection with the fear that the woman would exit the relationship (Browne, 1987).

Because of the high risk for death and injury, professionals involved with violent couples have a responsibility to understand and monitor factors which increase the risk of lethality. The following factors have been identified in studies by Browne (Browne, 1987), Walker (Walker, 1979), and Ewing (Ewing, 1987) as indicators of "higher lethality risks":

Physical Injuries: Physical injuries which are very serious and more frequent, especially if there has been a rapid escalation in the severity of battering are strong indicators.

Nature of the First Battering Incident: Homicide is more likely to occur between partners if the first battering incident involved life-threatening or severe violent acts or injuries.

Death Threats: Threats from either partner to kill themselves or others should be noted and monitored. These threats are of particular concern when they are detailed and when there appears to be the ability and intent to carry out the threat.

Presence of Weapons: Access to a weapon, especially a gun, is a major factor in predicting lethality.

Sexual Abuse: The frequency of sexual abuse of the battered woman seems to be greater in violent relationships which end in homicide.

Substance Abuse: The vast majority of homicides between intimate partners involve the abuse of alcohol or other drugs in one or both parties.

In addition to homicides and suicides, battered women fear physical injuries. Two-thirds of physical injuries requiring emergency medical treatment for women resulted from beatings by their husbands or boyfriends (Rosenbaum & O'Leary, 1981). The U.S. Attorney General has claimed that "battering is the leading cause of injuries to women—including rape, mugging, and automobile accidents" (Renzetti, 1988). The U.S. Surgeon General has stated that "battering is the single major health problem for women in this country" (Rosenbaum & O'Leary, 1981).

Physical Safety

The battered woman may literally have no safe place to go. Even though there are shelter programs throughout the nation (Roy, 1988) the woman may be unaware of these programs and the assistance which they provide. Even if she does receive shelter through a domestic violence program, this help is relatively short-term (usually 30 days) and she must consider longer term alternatives for herself and her children.

Economic Hardships

An obvious barrier to terminating violent relationships is the lack of money. The family's financial resources are usually strictly controlled by the abuser, so the woman may not have access to money (Ewing,

1987). Given the traditional values of these women, many have few job skills. When ordered by the court to provide child support or other forms of support to the family, the batterer may refuse and use the economic pressure as a way to force his wife to return or to persuade the children to live with him.

Lack of Support

Social isolation to which battered women are subjected by their abusers often narrows available support. Normally, women might turn to family or friends for financial and emotional support, at least temporarily. Unfortunately, when a battered woman turns to family or friends, they may blame her for the abuse she has suffered, and, even if they do believe her, they may advise her to stay in the relationship for the sake of the children and just try harder to be a better wife (Labell, 1979).

Ewing identified the typical response from helping agents as the most significant barrier to battered women seeking help (Ewing, 1987). Law enforcement officers, attorneys, judges, physicians, clergy and human service providers often fail to identify and address domestic violence. Schmidt (Roy, 1978) and Wilsnack (Schmidt, 1989, January) indicated similar gatekeeping barriers for alcoholic women.

ALCOHOL, DRUGS AND SPOUSAL VIOLENCE

Problem drinking in many physically abusive males has been documented by a number of researchers. Roy reported that 85% of battered women using a crisis hotline reported that their batterers abused alcohol (Seligman, 1975). For battered women seeking shelter, Labell reported that 72% of their batterers had drinking problems and 28.9% had drug problems (Labell, 1979). Fojtik found that 65% of the batterers abused alcohol and that two-thirds of assaults were alcohol-related (U.S. Attorney General, 1984). Walker indicated that alcohol was involved in over half of all battering incidents (Walker, 1979). Research conducted by Rosenbaum and O'Leary (Rosenbaum & O'Leary, 1981) and Van Hasselt, et al. (Van Hasselt et al., 1985) which specifically assessed alcohol consumption of abusive couples found significantly higher levels of alcohol use in physically abusive males. In violent lesbian relationships, Renzetti reported that 35% of batterers had been under the influence of drugs or alcohol at the time of battering incidents (Renzetti, 1988). Browne stated that abusive men with severe alcohol or drug problems abused their partners both when drunk and sober, were more violent frequently, and produced more serious injuries than did batterers without histories of substance abuse (Browne, 1987). Walker found that injuries were more severe in many cases when the batterer was intoxicated (Walker, 1979).

While each of these researchers emphasized that there was no evidence that alcohol causes battering, it is clear that continued substance abuse presents a tremendous barrier to stopping violent behavior. Violent acts committed while the batterer was intoxicated are often minimized by both the battered woman and gatekeeping systems. This failure to hold chemically dependent abusers accountable for their actions reinforces both addictive and violent behavior.

Studies reporting high rates of alcohol use among abusers found that the battered women fell within the normal range for alcohol use among women (Walker, 1979; Ewing, 1987; Browne, 1987; Labell, 1979; Rosenbaum and O'Leary, 1981; Van Hasselt, et al., 1985; Renzetti, 1988). Eberle found that the battered women who use alcohol are generally older than those who do not (Eberle, 1982). Walker noted that battered women who reported the heaviest drinking pattern for themselves were in relationships with men who also abused alcohol (Walker, 1979). In Roy's study of children from violent homes, the children reported that 41% of fathers had a drinking problem and that 24% of the mothers had a drinking problem to "forget the beatings" (Weisenfluh & Aniolkoski, 1988, May). Women receive twice as many prescription drugs as men. Weisenfluh reported that 25% of battered women compared to 10% of non-battered women receive mood-altering prescription drugs from emergency room physicians (Wetzel & Ross, 1983).

The chemically dependent battered woman faces special dangers. Friedman stated that these women are more likely to be physically damaged by the violence because the chemicals make them less able to escape violence and less aware of or likely to respond to injuries (Wilsnack, 1989, January). If she seeks help the chemically dependent battered woman may be rejected at domestic violence shelters because of her chemical abuse, and chemical dependency programs may fail to identify violence as a treatment issue (Wilsnack, 1989, January). These women face the double stigma of being labeled an addicted woman and a battered woman.

Spousal violence creates special problems for children. One-third of all American children observe their parents hit each other (Browne, 1987). In at least half of all homes where the wife is abused, the children are also abused (Eberle, 1982). Even when children are not physically abused, the abuser's violence toward his mate tends to cause a variety of problems for their children.

One of the most alarming findings in Roy's study was that 85% of children from violent homes admitted to a drinking problem which started as early as age eleven (Weisenfluh & Aniolkoski, 1988, May). Over 50% of these children had used marijuana or methamphetamines, and 10% were habitual users (Weisenfluh & Aniolkoski, 1988, May). Drugs and alcohol were used by the children to gain self-respect with peers and to "take the edge off having to face violence between their parents" (Weisenfluh & Aniolkoski, 1988, May).

DISCUSSION

In recent years, coexistence of chemical dependency and domestic violence has received increasing interest from both researchers and human service professionals. Clearly a need exists for education of professionals in both the chemical dependency and domestic violence fields so that effective service strategies can be developed and implemented. An essential part of a comprehensive service system is the education of community gatekeepers about the special needs of families impacted by both chemical dependency and family violence. Finally, prevention and early intervention services must be developed for children from these homes in order to break the intergenerational cycles of chemical dependency and domestic violence. Additional empirical research is needed in order to determine effectiveness of services and to understand the causes, manifestations, characteristics and connections between chemical abuse and domestic violence.

REFERENCES

Browne, A. (1987). *When Battered Women Kill.* McMillian, New York, NY

Dutton, D., Painter, S. L. (1981). Traumatic bonding: The development of emotional attachments in battered women and other relationships of intermittent abuse. *Victimology,* 6, 13–23.

Eberle, P., (1982). Alcohol abusers and non-users: A discriminate junctional analysis. *Journal of Health and Social Behavior,* 23, 260.

Ewing, P., (1987). *Battered Women Who Kill.* Lexington, Lexington, MA.

Fojtik, K. M. (1977–78). The NOW domestic violence project. *Victimology,* 2, 653–657.

Friedman, C., (1988 Summer). Alcohol and other drug abuse in the lives of battered women and their children. *NCADV Voice,* Washington, D.C.

Gelles, R., Straus, M. (1988). *Intimate Violence.* Simon & Schuster, New York, NY.

Labell, L. S. (1979). Wife Abuse: A Sociological Study of Battered Women and Their Mates. *Victimology,* 4, 258–267.

National Coalition Against Domestic Violence national toll-free hotline 1-800-333-7233.

Peachey, R. (1989, Winter). National estimates and facts about domestic violence. *NCADV Voice,* 12.

Renzetti, C. M. (1988). Violence in lesbian relationships: A preliminary analysis of causal factors. *Journal of Interpersonal Violence,* 4, 381–399.

Rosenbaum, A., O'Leary, K. D., (1981). Marital Violence: Characteristics of Abusive Couples. *Journal of Consulting and Clinical Psychology,* 49, 63–71.

Roy, M. (1988). *Children in the Crossfire: Violence in the Home—How Does It Affect Our Children?,* Health Communications, Deerfield Beach, FL.

Roy, M., (1978). *Battered Women: A Psychological Study,* VanNostrand Rinehold, New York, NY.

Schmidt, L., (1989 January). The Alcoholic woman in the 1980's. *The Counselor,* 13.

Seligman, M., (1975). *Helplessness: On Depression, Development and Death.* W. H. Freeman, San Francisco, CA.

U.S. Surgeon General (1984). *U.S. Surgeon General's Report on Family Violence.* Washington, D.C., U.S. Government Printing Office.

U.S. Attorney General, (1984). *Attorney General's Task Force on Family Violence: Final Report.* Washington, D.C., U.S. Government Printing Office.

U.S. Department of Justice (1986). *Criminal Victimization in the United States,* (a national crime survey report: NCJ-11-456). Washington, D.C., U.S. Government Printing Office.

Van Hasselt, V. B., Morrison, R. L., Bellack, A. S. (1985). Alcohol use in wife abusers and their spouses. *Addictive Behaviors,* 10, 127–135.

Walker, L. (1979). *The Battered Woman.* Harper and Row, New York, NY.

Walker, L. (1986). *The Battered Woman Syndrome.* Harper and Row, New York, NY.

Weisenfluh, S., Aniolkoski, C., (1988 May). Substance Abuse Among Battered Women: Developing a Model for Identification and Intervention of Domestic Violence Victims. Paper presented at the Alcohol and Drug Problems Association Second National Conference on Women's Issues, Columbus, OH.

Wetzel & Ross (1983). Psychological and social ramifications of battering: observations leading to a counseling methodology for victims of domestic violence. *Personnel and Guidance.*

Wilsnack, S., (1989 January). Women at High Risk for Alcohol Abuse. *The Counselor,* 16–17.

www.ingramcontent.com/pod-product-compliance
Lightning Source LLC
Chambersburg PA
CBHW050805160426
43192CB00010B/1645